THE GOD IMPERATIVE

Why We Need Faith to Safeguard Reason, Science and Liberty

DAVE DENTEL

Half A League Onward Press
Stephens City, Va.

The God Imperative: Why We Need Faith
to Safeguard Reason, Science and Liberty

Half A League Onward Press
Stephens City, Va.
www.halfaleagueonward.com

Edited by Kate Dentel

On the cover: Storm clouds darken the sky above an abandoned church in the Netherlands.
Photo by Jascha Hosk, stock.xchng Used by permission.

ISBN 978-0-615-46129-8

Library of Congress Control No. 2011905260

Contents

Thanks to

Phillip Johnson and Paul Johnson for inspiration

And to Kate for that and everything else

Introduction

When I finished graduate school and launched a career in newspapers, I braced myself for what I thought would be a lonely sojourn in a godless enclave. To my delight I discovered I was mistaken. Not all journalists, it turns out, are liberal pagans. In various newsrooms I even met up with a good number of fellow Christians, though most, it seemed, felt like me that it was prudent to pick their battles carefully. As for the liberal pagans I did encounter, imagine my chagrin when I learned even they can be human, warm, and generally kind.

There were exceptions. Some of the people I worked with acted as though it were a personal duty to uphold the cliché of the leftist ideologue. These folk championed the usual abstractions — tolerance, diversity, social justice — then to prove just how devoted they were to the cause openly despised any detractors. Most of the time this disparity between what they insisted they believed and how they behaved was simply amusing — until it metastasized into a moral disconnect that could be truly disturbing.

Some examples: About a week after George W. Bush's re-election, I decided to approach my boss to ask if I could schedule time to meet with a small group from my church. Minutes before my opportunity arrived, something provoked him and he started to rant about the religious conservatives who re-elected Bush. He said they were "whackjobs." Another colleague agreed. "Mini-Bushes," she said.

I brushed it off as post-election sour grapes. After all, these two weren't nearly as strident as another colleague — a self-styled Buddhist flower child — who said during campaign season that she wished Bush's motorcade would pass through town so she could throw rocks at his limo. (Maybe it was just the pot talking — and I don't mean the one that called the kettle black.)

Then there was the rookie reporter who had just returned from covering a pro-war, support-the-troops rally, no doubt replete with flag-waving and jingoistic calls of God Bless America. Apparently she'd never seen anything like it at the Ivy League school she attended. She called the rally-goers fascists.

I end with one of my personal favorites. Years earlier at another newspaper, a colleague of mine — a Green Party activist — couldn't help but comment on a speech the Roman Catholic pontiff had just delivered about birth control. "Doesn't the pope know," she demanded, her hands resting on her visibly pregnant tummy, "that the world is overpopulated?"

Recall that these are folks who consider themselves professionally objective and the truth behind the next point becomes painfully obvious. Journalists as individuals may rise above the left-wing stereotype, but collectively the news they produce usually conforms to the typical liberal point of view. There's a reason for this: Journalists tend to be liberal and irreligious. They are schooled by professors who tend to be liberal and irreligious. And they labor in a highly competitive industry that tends to reward reporting that is liberal and irreligious.

As op-ed columnist John Leo puts it, most news coverage automatically gets shaped to fit certain storylines, or templates. And these storylines inevitably reflect the values of those who fashion them. This is not the result of conspiracy. It is simply the way the institution of journalism functions.

For example, nowadays the media almost always depicts environmentalism as positive and progressive. Heavy industry, on the other hand, is usually portrayed as archaic and dangerous. Thus legislation blithely mandating the sales of compact fluorescent lightbulbs easily earns the labels "earth-friendly" and "green," despite the fact these products contain toxic heavy metals and require special means of disposal. Meanwhile, an oil company's announcement that it plans to drill in untapped areas of the domestic reserve automatically triggers the questioning, cautious storytelling template reserved for controversies.

To digress: I recall being annoyed by a few newsroom colleagues who constantly harped on the sins of certain allegedly greedy industrialists, usually referred to by disparaging sobriquets such as "Big Tobacco" and "Big Oil." In response, I facetiously began denouncing the evils of "Big Corn Syrup" (it really is in just about everything you eat or drink). Imagine my surprise when not too long ago — once obesity had become the nanny culture's latest bete noire — the corn syrup lobby actually began taking out ads in defense of its product!

Back to the point. One particularly unfortunate example of journalism's institutional bias in framing issues is in its treatment of evolution, or more specifically, its treatment of those who object to Darwinism's status as the de facto official creation myth of American public education. Indeed, science and the theory of how living things evolve may have changed over time, but the way in which mainstream journalism is likely to report on dissent concerning that theory hasn't altered much since H.L. Mencken covered the Scopes Monkey Trial.

In his reporting of the iconic 1925 court case, Mencken depicted opponents of evolution as ignorant yokels whose religious fanaticism not only blinded them to the truths of science, but made them bigots. Fast forward to the Dover School Board trial in 2005 and the characterization by the press is remarkably similar. For instance, in a documentary on the federal trial by the public television science program *Nova*, the narrator describes Dover School Board members — and their supporters — who objected to the teaching of Darwinism as "creationists" who "reject much of modern science."

To digress yet again: This accusation always struck me as more than a little odd. Frankly, it's not as though questioning the efficacy of natural selection in creating new life forms is really all that unorthodox. After all, though Darwin himself set out to explain the origin of species, the fact remains that scientists can't necessarily explain what a species is, agree on how various life forms are related or how they should be categorized. So it's not as if doubting whether an ancient mouse-like mammal really could evolve into a humpback whale automatically is going to undermine a person's confidence in things like Newtonian physics or the periodic table of the elements.

Still, given that this less than flattering treatment of dissenters from Darwinism (especially religious ones) is standard fare in the mainstream press, it's remarkable that during the Dover trial my editors approved of the commentary I provided while on staff at one of the local dailies. What I wrote then certainly must have seemed unusual. It was informed, reasonable and professional. But it supported the emerging theory of intelligent design and questioned evolution — especially in the way some of its most ardent adherents use it to attack religious belief. I know my op-ed columns struck a chord, because I received numerous responses, not only from locals but from readers as far away as Australia. (And once, after shamelessly Googling myself, I saw one pro-evolution blog refer to me as my newspaper's "in-house creationist troll.")

Less than remarkable was the outcome of the trial, which predictably affirmed Darwinism as the only explanation for the development of life permissible in public school science class. In part, the result was provoked by the actions of certain school board members, who engaged in what can only politely be called shenanigans while trying to disguise their religious motivation for challenging the teaching of evolution. Though their religious objections to Darwinism were understandable, as an impetus for introducing new science curriculum they did not pass legal muster.

It should be pointed out that technically the court's ruling applied only to the Dover public school district. But it is certain that the effects of the federal court decision were felt far beyond the small jurisdiction in southeast Pennsylvania.

I for one believe the Dover trial emboldened militant atheism and fueled its adherents' antagonism toward religion, not to mention their disdain for public policy inspired by religion. In books, on the lecture circuit, in the press, and in such unexpected places as Christmas billboards (even the Grinch wasn't that heartless), atheists in recent years have escalated the call for converts in an effort to finally displace faith — particularly Christian faith — as a major source of cultural influence in the West. And in disparaging theism, these new atheists offer in its place culture-forming narratives based in great part on evolutionary science. Life, human thought, the principles of civil society — all are to be traced to a genesis in random mutation acted upon by natural selection. From this indifferent beginning, atheists are then free to fashion a godless reality where moral truth is a human invention and the right thing to do can be determined only by the kind of careful scientific research that seeks to prove apes typing randomly on computer keyboards really can produce the works of Shakespeare. (Someone actually tried this; all the apes produced was gibberish and copious amounts of dung.)

And this escalated attack on faith has had its effect. To begin with a personal example, while participating in a religious event at a community college in Virginia I was accosted by a student who wanted to provoke an impromptu debate on the merits of science versus faith. Anticipating a barrage of the usual sophomoric materialist objections, I avoided a lengthy discussion since I did not feel the occasion lent itself to such an exchange. Later, however, I found the student had cornered a colleague of mine and was peppering him with demands for physical proof regarding the credibility of revealed religion.

Again, much of what the student had to say was predictable. He chiefly parroted the new atheists. When I told him this he merely repeated another new atheist dodge in insisting that atheism isn't new and that, after all, George Washington was an atheist, or at least a deist, a misrepresentation that completely missed the point. (A good deal of our time was spent talking past each other, which was why I had hoped to avoid the conversation in the first place.)

Perturbing as it was to hear the anticipated anti-religious palaver, however, what proved truly disturbing was when the student went on to attack religion with assertions wholly disconnected from reality. In an attempt to make a point about the need for divine justice, for instance, I invoked Hitler, which is nearly always a mistake owing to the fact that the Fuhrer has been turned into such a caricature of evil that he can be employed to discredit just about anything. (A new surtax on dessert cheeses?

That's just like Hitler!) Imagine my consternation when the student rebutted my argument by asserting that Hitler was a Christian who persecuted the Jews because he felt they were collectively responsible for the death of Christ. From there the student tried to celebrate the triumph of science by employing Pascal's Wager, which he said had something to do with celestial teapots orbiting alien stars. Of course, my sad conclusion was that social critic Neil Postman was right when he said that our society's blind regard for science and the fruits of technology has produced a generation whose members can no longer distinguish fact from fantasy.

Then there's the broader evidence as to the influence of the new atheism. One poll shows the number of Americans who call themselves Christians has fallen 10 percentage points since 1990. In that same time frame the number of self-described atheists in America has increased from 1 million to about 3.6 million. It is this trend that prompted one national news weekly magazine to ask whether we have reached the end of Christian America, at least in terms of Christianity's influence in mainstream politics.

Of course, as previously noted, the collapse of Western Christendom and the rise of a new culture shaped by secular reason represents the very thing for which the new atheists are striving. And with the aid of the evolution narrative and the liberal-leaning media they may indeed achieve their goal. What we have to ask ourselves is this: How far will society slide into the moral abyss if they do?

Chapter One

A New Crusade Against Faith

Terror, Materialism and the Battle for Western Civilization

"Violent, irrational, intolerant, allied to racism and tribalism and bigotry, invested in ignorance and hostile to free inquiry, contemptuous of women and coercive toward children: organized religion ought to have a great deal on its conscience."

– Christopher Hitchens, *God is Not Great: How Religion Poisons Everything*[1]

"Should we bash religion with a crowbar or just a baseball bat?"

– Melvin J. Konner, anthropologist criticizing fellow scientists during a symposium on science and religion[2]

Every revolution needs a rallying cry, and every crusade requires some banner under which its holy warriors can march. Injustice, catastrophe, scandal — these common woes can also serve as tools for defining enemies, characterizing threats, winning converts and agitating for change. For leaders of one of this young century's most ardent — and unsettling — social and intellectual reform movements, just such a revolutionary icon was forged on September 11, 2001.

What happened that day is well established. Nineteen young men, all Middle-eastern and all Muslims, hijacked four commercial airliners from airports in the Eastern United States. Two of the planes were guided into the twin towers of New York's World Trade Center, which collapsed shortly afterward. A third was flown into the Pentagon, the U.S. military office complex in Northern Virginia. The fourth crashed in a field near Shanksville,

Pennsylvania after passengers and crew attempted to retake control of the aircraft. Two thousand, nine hundred and thirty-seven people were killed in the attacks, which ultimately yielded more death and destruction when, in retaliation, the United States and its allies invaded Afghanistan and Iraq to flush the terrorists from their bases of operation.[3]

> **Key Concepts**
>
> The new atheists claim:
> ❖ Religion is a dangerous lie that inspires its adherents to commit violence.
> ❖ Humans should look to science alone for truth.
> ❖ Since all morality is manmade, only secularists with the proper scientific outlook should be trusted to shape moral law.

These facts alone are enough to appall any rational conscience. The audacity of the terrorists and the carnage they perpetrated present an insufferable affront to anyone who cherishes life and the merits of democracy, and it is more than reasonable that those who feel most offended should desire retributive justice.

And yet what goads the leaders of this new crusade is not simply what the terrorists did, but why. The notion that 19 hijackers committed mass atrocities in the name of their god, that they felt compelled to destroy as a fanatic expression of their religious faith, mystifies and frightens these would-be reformers — who have no god greater than themselves.

The reason for this visceral reaction is simple: The leaders of the new crusade are atheists. They not only reject the proposition that God exists and is concerned with the affairs of human beings, they insist that belief in God and the practice of religion is sheer folly and ignorance that can easily lead to malign acts. In the September 11 attacks, they claim they have the proof to support their dire accusations, and consequently have announced their intention to make religion unwelcome in liberal, democratic society — with the hope that someday theistic faith might be made to disappear entirely.

Not that questioning prevailing religious views, even to the point of atheism, represents anything new. One could even say that Adam set the precedent for his descendants to struggle with faith when he accepted the forbidden fruit from his mate in the garden of Eden.

Since then, in Western culture especially, disdaining the religious establishment for the sake of championing a new social order has become something of a tradition. Socrates, before he drank the hemlock, conceded he considered the Athenian pantheon an impotent, obsolete myth. Jean Jacques Rousseau's ideas helped the Enlightenment veer from law based on God-given rights to a human declaration of equality and fraternity as enforced by mob rule. The following century Friedrich Nietzsche gave the world both the fantastic concept of the amoral superman who had passed beyond his need for God and the very real, very tragic example of a godless philosopher who had lapsed irretrievably into madness.

Science Turns From God

But it was Charles Darwin, riding on the coattails of lesser-known proto-evolutionists, who finally provided agnostic-leaning intellectuals the tool to construct a comprehensive atheism. Darwin's principles of evolution — that all living things are the product of random mutation acted upon by natural selection — proposed a model of life origins that made God unnecessary. Darwin's principles appeared to be confirmed by scientists in other disciplines. Geologists and paleontologists declared the evidence pointed to an ancient, ever-changing world where for the greater part of its history human beings (the only creatures who recognize and worship supernatural beings) didn't exist. Cosmologists extrapolated the apparent great age of the universe and described its formation in terms of random evolution. Physicists puzzled over the bizarre behavior of sub-atomic particles and decided to express the way they function in terms of uncertainty.

In short, Darwin supplied science — especially the underlying philosophy supporting its methodology — with the ethos it needed to become at last inexorably secular. Since the rise of Darwinism, any sort of theistic rationale for where the universe and its inhabitants came from, or why things function they way they do, has been pushed so far to the periphery that it is no longer fashionable to invoke God even when considering first causes. A materialistic outlook now prevails to the point that most leading scientists confess to at least some degree of agnosticism.[4]

This is not to say that science, or even materialism, are in themselves bad things. Indeed, when considered in its most basic definition as a method for studying whatever phenomena we encounter (and most of these, as it happens, are material in form), science has proven one of the best things human beings ever devised, and has yielded tremendous successes. What concerns us, then, is not science as methodology, but science distorted by those who exploit its achievements in order to falsely elevate it above other means by which human beings seek to understand their surroundings. This is science as ideology.

Given this distortion, it should hardly come as a surprise that the views of leading scientists have consequently attained undue influence over society at large — even concerning matters of faith. Historian Paul Johnson notes the impact of science on prevailing ideas when he outlined how at the dawn of the twentieth century American and European intellectuals embraced the theories of Albert Einstein and Sigmund Freud at the expense of traditional Judeo-Christian mores. Never mind that they mistook relativity for relativism, or failed to see the gnostic nonsense underlying Freud's vague theories of the subconscious. The oracles of science had spoken, and, as Johnson explains, "the scientific genius impinges on humanity, for good or ill, far more than any statesman or warlord."[5]

In fact, science has since attained such ascendancy that the only creation myth generally tolerated in Western state-funded schools is an utterly materialistic one. Even combining the words "creation" and "science" in an educational setting is, for a certainty, to provoke institutional ridicule.

As a consequence, the assault on a Judeo-Christian framework for understanding reality has reached a historical zenith. Few Christians today dare display the temerity of James Ussher, a seventeenth-century Anglican bishop who concluded, based solely on the Bible and classical history, that God created the world on Sunday, October 23, 4004 B.C.[6] Modern science insists that the physical evidence shows such a declaration to be utterly false, if not absurd. And modern science, thanks to its innumerable practical successes, has attained a position of authority that is nigh unassailable.

> **Key Quote**
>
> Darwin supplied science with the ethos it needed to become inexorably secular. And modern science, thanks to its innumerable practical successes, has attained a position of authority that is nigh unassailable.

Faced with such intellectual denigration, many contemporary religious thinkers have come to accept the sop thrown them by condescending scientists — namely, the notion that religion still has something to contribute to the storehouse of human knowledge as long as that something doesn't presume to be actual facts. Oddly, this idea that religion and science should be segregated into separate realms of learning, or as paleontologist Stephen Jay Gould suggested, defined as "Non-Overlapping Magesteria," is often put forward as being beneficial to religion.[7]

Indeed, Gould imagined that establishing recognized boundaries would help engender in others what he declared was his own great respect for religion, particularly as a means for understanding ethics and morality. What he affirmed as respect, however, was unfortunately undermined both by his own agnosticism and his granting primacy of science over scripture. Gould's declaration that religion's importance lies in its authority over morals appeared similarly dubious when he insisted that in order to "formulate a moral theory under the magesterium of ethics and meaning" humans "need not invoke religion at all."[8]

This is not to say Gould was completely off base. The notion that science and religion should remain confined to separate compartments does serve some usefulness — at least in an academic sense. Most people, for instance, would agree that scientists confronted by particularly baffling research problems should pursue answers that conform to observations within nature, and not simply shrug and say, "Oh well, God just made it that way." And scientists operating within proper ethical guidelines should not be constrained simply because their findings appear to threaten cherished religious doctrines concerning some aspect of nature. On the other hand, scientists should not be permitted to claim that their research into the workings of nature nullifies truths pertaining to the supernatural — truths which by definition lie clearly outside the scope of science.

When Faith Won't Die

But as useful as it may be for political or academic purposes, in the long run any dichotomy imposed upon religion and science must be judged an artificial separation. Because ultimately, what is true about such fundamental things as why the cosmos exists or how human beings are supposed to live encompasses not just science and religion, but every branch of human learning we could name.

To be accepted as viable, then, a worldview must exhibit a coherence that breaches artificial intellectual barriers — particularly those erected between science and faith. Many ardently religious people recognize this, as do many outspoken atheists.

After all, the September 11 terrorists didn't stop to think about whether or not they should employ the scientific principles of aeronautics to help them commit religious violence. They didn't ask if human behavior studies supported the ethics of crashing hijacked airliners into buildings, or whether their own deaths for Allah's sake were undermined by scientific evidence that calls into question whether any deity exists. They simply believed that their god required them to kill and acted on that belief.

Which brings us to the present crisis.

Given the obvious evil that religious extremism can lead to, and given the logical conclusions drawn from prevailing scientific philosophy, many leading atheists are questioning why religion should be tolerated at all. This is not a mild, academic querying. It is a movement stoked with evangelical fervor, a movement that is resonating with the public. Among themselves, in the popular media, and in a wave of best-selling books, atheists have questioned the value of religious faith and have challenged believers to prove whether religion offers anything true, or good.

The raptures to which this agnostic awakening have soared was vividly described by *New York Times* reporter George Johnson in his coverage of a symposium on science and religion held in November 2006. As Johnson characterizes it, the elites who gathered at California's Salk Institute for Biological Studies quickly rejected any pretense of seeking dialogue with theism, calling instead for a renewed commitment to changing the world through science.

"We should let the success of the religious formula guide us," research scientist Carolyn Porco declared, offering a sort of backhanded compliment to colleagues of faith — who largely avoided the conference. "Let's teach our children from a very young age about the story of the universe and its incredible richness and beauty. It is already so much more glorious and awesome — and even comforting — than anything offered by any scripture or God concept I know."

Steven Weinberg, a Nobel laureate in physics, concurred.

"Anything that we scientists can do to weaken the hold of religion," he said, "should be done and may in the end be our greatest contribution to civilization."

And just in case anyone had failed to see the urgent need for finally muzzling theism, best-selling atheist author Sam Harris delivered a rousing altar call.

"I don't know how many more engineers and architects need to fly planes into our buildings before we realize that this is not merely a matter of lack of education or economic despair," he warned.[9]

Despite the rather absurd image of top scientists thumping the podium like so many camp meeting revivalists, they and fellow militant atheists have raised questions that demand to be answered. Is it reasonable to believe God exists? Is religion true, or a malicious farce? Can reason alone, without regard to supernatural agents, religious mores or even long-standing taboos, lead us to peace, prosperity, and harmony with our fellow creatures?

What's more, the atheists' queries represent something deeper than anxiety during unsettling times. They also represent the logical culmination of modern scientific thought, centered particularly around Darwinism. If the cosmos is indeed the product of mere chance and blind law, why shouldn't we conclude that God is a lie and man very much alone in his quest to find purpose, peace and fulfillment in a universe completely indifferent to his struggles? Why shouldn't we cast off religion and look for something better?

Evidence Demands Another Look

Because the conclusion that science leads away from God is a false one. Taken as a whole, the evidence unfolding from our observations of the physical universe, from the fine-tuned forces that hold it together to the complex genetic code residing in every living creature, points away from chance and toward meaning and purpose — in a word, intelligence. Continued inquiry into the history, character and nature of humans only seems to confirm the fact that we are special creatures, but highly flawed. We need help, and it is more than reasonable to hope that help can come from a God big enough not only to create the cosmos but to sustain it and ultimately redeem it.

Chapter Two

Godless Hubris

How Atheists Distort Science to Undermine Religious Values

"Clearly, it must be possible to bring reason, spirituality, and ethics together in our thinking about the world. This would be the beginning of a rational approach to our deepest personal concerns. It would also be the end of faith."

– Sam Harris, *The End of Faith: Religion, Terror and the Future of Reason* [1]

"Reason itself is a matter of faith. It is an act of faith to assert that our thoughts have any relation to reality at all."

– G.K. Chesterton, *Orthodoxy* [2]

As strange as it may seem, a rational path back to faith in a creator and redeemer actually begins with the claims of today's leading atheists.

As we have already noted, the impetus for their claims is not so far out of bounds. They express deep confidence — awe even — in the notion that nature displays great complexity and apparent order. But they're also deeply dissatisfied with what is so obviously wrong with nature.

This paradox of wonder and fear, delight and disgust is common. Quite uncommon, on the other hand, is what the new unbelievers seek to make of this paradoxical outlook concerning creation. In their view, awe and wonder should be reserved for the indifferent mechanism they claim more or less accidentally made these emotions possible. And all ire should be directed at the false concept that seeks to direct credit for the cosmos toward some supernatural entity.

The irony, of course, is that if Darwinian forces did indeed bring about our reality, then religion is also merely the product of chance and selection and is no more worthy of being hated because of what it purports than the sky is because it's blue. But then, as we'll see, religion apparently rates a special kind of hatred because it is intrinsically related to other human characteristics that are very problematic to Darwinists — consciousness and free will.

Key Concepts

The new atheists differ from their godless predecessors in novel but disturbing ways:

- ❖ They praise democracy and civil liberties even though they don't believe in protecting religion.
- ❖ They say the wonders of the cosmos can evoke a quasi-mystical response, but this claim is mainly an excuse for making science into a religion.
- ❖ They blind people to truths about divine justice and grace.

And there is no doubt that many of today's leading atheists are not of the Enlightenment fop variety. They actively hate religion. Indeed, the most militant opponents of faith consider it the leading cause of all the blight and mayhem in human history.

Consider, for example, how Oxford professor and atheist apologist Richard Dawkins disdains religion while simultaneously rhapsodizing about his vision of utopia — a faith-free world. In prefacing his best-selling book *The God Delusion*, Dawkins invites readers to:

> "Imagine, with John Lennon, a world with no religion. Imagine no suicide bombers, no 9/11, no Crusades, no witch-hunts, no Gunpowder Plot, no Indian partition, no Israeli/Palestinian wars, no Serb/Croat/Muslim massacres, no persecution of Jews as 'Christ-killers,' no Northern Ireland 'trouble', no 'honour killings', no shiny-suited, bouffant-haired televangelists fleecing gullible people of their money ('God wants you to give till it hurts')."[3]

Vitriol aside, on one level Dawkins' litany of the failures of religion is quite fair. Injustices committed in the name of faith are a matter of record; no one can dispute that these things have occurred. What can be disputed, however, are the circumstances connected with these very human failings, things such as motive, provocation, miscommunication and in many cases, outright mistaken belief. These finer points Dawkins chooses not to dwell on, however, since it appears his main purpose is to cast religion as the scapegoat for all of history's ills.

Again, this general railing against faith is not so profound or unusual, until one considers the reasons behind the rants. Because it is there — in the heady mix of scientific hubris and political cant underpinning contemporary atheist philosophy — that the new unbelievers' crusade reveals how it is both novel and dangerous. Consider:

■ That atheist scientists exploit both the material blessings obtained through their research, and the position of authority these blessings afford them, in making a moral judgment against what has traditionally been considered the source of morality.

■ How the new unbelievers tout science not just as a tool for curbing ignorance and superstition, but as a comprehensive belief system in itself.

■ That some atheist advocates display few qualms in the notion of using political power to endorse a materialistic orthodoxy and to quash religion.

■ Realize, too, that in the age of suicide bombers and weapons of mass destruction, some of the most militant atheists warn that tolerating theistic faith in any form is too perilous a luxury.

As Harris writes, "I hope to show that the very idea of religious tolerance — born of the notion that every human being should be free to believe whatever he wants about God — is one of the principal forces driving us toward the abyss." He adds: "We have been slow to recognize the degree to which religious faith perpetuates man's inhumanity to man."[4]

So there is fear here, certainly. But the fear is misdirected, if atheists, agnostics and other assorted disparagers of faith think they can simply attribute all of humanity's woes to belief in a God who made us and loves us. Their confidence in science is misplaced, as well, if they believe that by simply understanding the mechanisms of the cosmos we can somehow escape the perils of sin and death that pervade our reality — perils that virtually define what it is to be human.

Ironically, the new unbelievers could grasp these truths if they only approached their study of our universe in the honest, disinterested manner that they claim to put forth. As it is, their science and philosophy only result in profound paradoxes — enigmas that should point them away from the indifference they claim to have discovered and toward the ultimate source of meaning itself.

"Religious non-Believer"

Before atheist charges can be rebutted, however, they need to be more closely examined. But first, a word of explanation: It is convenient rhetorically to employ the term "atheist" for current outspoken foes of religion such as Dawkins, Harris, journalist Christopher Hitchens and Darwinian philosopher Daniel Dennett. In fairness, though, it should be pointed out that they and others aligned with their ideas claim various degrees of disbelief.

Dennett, for instance, prefers to consider himself a "bright,"[5] a euphemism he apparently thinks holds promise as one of the aforementioned banners under which his fellow travelers can unite. If pressed, he would probably accept the label "agnostic," a term which Dawkins eschews.[6] Nevertheless, both claim to be open to the idea that there is more to existence than the mere physical. Dawkins goes so far as to label himself a "deeply religious non-believer." He explains: "A quasi-mystical response to nature and the universe is common among scientists and rationalists. It has no connection with supernatural belief."[7] Harris, as well, offers vague support for the notion that spirituality can provide insight into truth. It's as though they are hedging against their expectation that science can achieve anything. They don't want to be found completely wanting in faith on the odd chance they should be proven wrong.

More instructive is what these four obviously hold in common, because it defines what I mean when I discuss members of the new atheist crusade. This is their creed: With uplifted fists they vociferously deny the existence of God; they all adhere passionately to Darwinism, and they all consider religion a dangerous lie.

Sure of Nothing

All three ideas are closely interwoven. And based as they are on a faulty understanding of reality, they nonetheless represent a relatively logical progression.

To begin with, the new atheists dismiss all arguments for the existence of God as weak, irrational, or irrelevant. Classical philosophical approaches for proving God exists, such as the ontological argument, do nothing for them. Dennett goes so far as to call these "largely wasted effort."[8] As for the argument from design, Dawkins famously insists that though the Earth — especially the life it contains — certainly looks as if it were designed, Darwinism proves that it is not.[9]

Scripture they disparage as vacillating between nebulous half-truth and malicious nonsense. Personal claims concerning supernatural revelation they equate with lunacy. Miracles, they insist, are simply impossible.

The new unbelievers can be sure of themselves in dismissing the supernatural because they have found something they are certain makes belief in it quite unnecessary — modern science.

Part of this certainty, of course, lies in semantics. By definition science deals with physical reality — things that can be tangibly observed, analyzed, quantified and ultimately understood in terms of predictable reactions and outcomes. In short, as Gould argued, science deals with facts. And though it is true a good many intangible things appear to lie outside this sphere of understanding, and that the origin and nature of these things may be the subject of much conjecture, according to science their very existence can never be confirmed. In this way materialist philosophy and science become inextricably intertwined.

A materialist philosophy, however, is fairly worthless without some mechanism for explicating the material universe. And once again, the mechanism the new unbelievers put their faith in is Darwinism.

Understand, though, that Darwinism in this sense is not merely a tool for explaining how new species form, or how slow, random change supposedly produces biological features such as eyes, lungs and wings. Particularly since the rise of gene studies in the mid-twentieth century, Darwinists have increasingly attempted to apply the principles of replication, mutation and selection to the most obscure characteristics of living things. Consciousness, behavior, and even morality and ethics have come under the purview of Darwinian researchers. Dawkins goes so far as to propose that ideas and knowledge — what he termed "memes" — exist as independent entities that are also the product of aimless evolution.[10] His theory has been embraced and expanded upon by several others, including Dennett.

In light of such an all-encompassing point of view, it is easy to see how the new unbelievers can claim that religion also falls under the bailiwick of science. But the fact that scientists study religion is not to be construed as an endorsement of religious teachings, which atheists insist remain utterly false. Instead, religion itself is considered an actual, albeit bizarre, manifestation of evolution. And as such, it is simply natural phenomenon, in the total scope of things no more profound than a moth immolating itself in a candle flame.[11]

Indeed, what purpose religion serves, and whether it offers anything of value, remains a matter of debate. Some atheists grudgingly allow that religion may provide a basis for altruism — another concept that proves troublesome in the survival-of-the-fittest world of Darwinism. The more ardent atheists, however, assert that religion can best be explained as something alien and viral.

Dawkins muses, in typically nebulous fashion, that in humans, "the religion behavior may be a misfiring, an unfortunate byproduct of an underlying psychological propensity which in other circumstances is, or once was, useful."[12]

Dennett concurs. In fact, he insists the moment has come for humanity to determine once and for all whether theistic faith offers any benefits — through scientific inquiry, of course.

"It is high time that we subject religion as a global phenomenon to the most intensive multidisciplinary research we can muster, calling on the best minds on the planet," Dennett declares.[13] (One presumes his mind heads the list, at least, ahem, in his own mind.) He adds, rather ominously, "Only when we can frame a comprehensive view of the many aspects of religion can we formulate defensible policies for how to respond to religions in the future."[14]

False Distinctions

Given such presumptuousness, it would be easy to dismiss Dawkins, Dennett and his ilk by lumping them in with the all-too-familiar atheist cranks of the not-too-distant past. Easy, except for the fact that the new unbelievers have also taken great pains to distinguish themselves from certain unappealing philosophical forebears. And these distinctions deserve our notice, if for no other reason than to expose them as new spin on old irrationality.

The new atheists, for instance, are neither Marxists nor anarchists. They heap praise on aspects of Western democracy, including representational government, rule of law and civil liberties — at least, to a point. They conveniently forget the Judeo-Christian ethos from which Western government stems. In other ways their recognition of the blessings of democracy seems less a measure of gratitude than a fearfulness that the prosperity and personal freedom they so enjoy may be quickly taken away.

They certainly aren't above sanctioning a little judicious tyranny, so long as it serves their own interests.

Nor are the new atheists libertine freethinkers — transparent hedonists whose attacks on public virtue barely disguise their own yearnings to sate every personal vice. On the contrary, they declare that in a secular, liberal democracy an orthodox system of ethics is not only attainable, but imperative. Again, however, they leave plenty of wiggle room. Since the new atheists deny any moral absolutes, the ethics they forward tend not so much to restrict any specific behavior as always wrong, but instead provide parameters for privileged individuals to dabble in debauchery without propelling society at large into moral chaos.

Key Quote

In all too predictable fashion, the new atheists end their flow of materialist logic by calling for the establishment of a new secular priesthood and then nominating themselves for ordination.

Ultimately, though, the new atheists fail to distinguish themselves from so many godless thinkers who came before in one important respect. Starting from what they say is a position of fact, they build on supposed moral truth in order to overthrow existing standards and present themselves as the new arbiters of reality. If, indeed, as the new unbelievers insist, religion has from the beginning been a false hope, then to whom will peace-loving, right-thinking people turn for moral guidance in everything from personal manners to public policy?

As if the answer weren't painfully obvious. In all too predictable fashion, the new atheists end their flow of materialist logic by calling for the establishment of a new secular priesthood and then nominating themselves for ordination.

And yet their logic doesn't hold up. In constructing a worldview based on materialistic scientism, what they yield at last are not liberating truths but insurmountable paradoxes.

Consider the new atheists' position on good and evil. Despite a Darwinian outlook that says the cosmos is fundamentally indifferent, pointless even, they recognize evil as very real. They go so far as to insist that because there is so much evil in the world, no beneficent creator could possibly exist. Yet they base this denial of deity on the great sense of goodness they purport to hold within themselves — goodness they say is scandalized by the blatant evil for which no natural or supernatural being is responsible, but nonetheless exists. So the new unbelievers are left with the unpleasant fact that despite rejecting a divine creator, they still have to explain how any creation system they do ascribe to can be considered good in any sense if it produces evil things.

From this shaky foundation rises an even more precipitous theory of morality. Having demolished a divine source of moral law, they are left to construct a system of ethics based solely on human understanding, a practice which, as we have already noted, tends to yield less than satisfactory results. Science doesn't necessarily help in this regard. As we shall see, scientists who do try to construct a Darwinian model of morality, for instance, produce some very contradictory ideas since they cannot avoid struggling with evolution's determinist underpinnings. And generic scientific theories are easily perverted. Nazi race-laws and Stalinist social engineering attest to this. So without an objective standard, the new atheists' idea of right and wrong boils down to individuals acting according to what they believe to be their own best interests — which is very nearly the same thing as having no morality at all.

Lastly we come to the paradox of how the new atheists are able to arrive at any conclusion at all without invoking the very thing they claim does not exist — the supernatural. Indeed, the most ardent Darwinists among them insist on viewing the cosmos as the result of constant evolution, even down to the very thoughts and ideas that supposedly compete for dominance in the minds of human beings as memes. Religion, for instance, is purportedly just another meme. But what of evolution itself? If what the new unbelievers imply is true, then Darwinism is not just another transient idea, but an exception to constant change which serves as the overarching principle governing the cosmos. And so are other highly regarded cornerstones of science — mathematics, physics, genetics; in fact, if these concepts were not fixed it would be hard to imagine how any observer of nature would be able to make any sense of them at all. Yet, if laws such as those proposed by Darwin never change, that would place them outside the realm of nature, which we are told is constantly changing. And that would make them supernatural.

Ultimate Judge

As facile as this objection may seem, it nevertheless points out the danger behind the ideas and attitudes being forwarded by new atheists. In denouncing theistic religion as vain superstition, they claim to offer what Dennett calls " a better way" through rational inquiry, a sort of humanistic faith inspired by science. In actuality what they do is undermine the very things they say they champion.

By dogmatically limiting human understanding to what can be derived by materialistic inquiry alone, they shatter the rational basis for observation and experimentation that are the backbone of science. And without a rational basis for trusting what we observe to be true, individuals are much more likely to embrace what is irrational and absurd as it suits them. In a very similar way, by rejecting any notion of absolute right and wrong and relying instead on human reason for determining how we should live, the new atheists actually pave the way for immorality, cruelty and oppression.

But the biggest danger by far the new atheists pose is in obscuring the stakes involved in taking them at their word. If, as Blaise Pascal argued, those who live moral lives lose nothing if it turns out God does not exist, what does that say about the peril implicit in the opposite scenario? If God exists, then all arguments about whether or not he left enough proof of his existence become inconsequential. The only thing that will matter is how prepared we are when we stand before him in judgment.

Chapter Three

DOUBTS ABOUT DARWIN

Questioning The Agnosticism Implicit in Evolution

"The concept of evolution is the cornerstone of biology, because it links all fields of life studies into a unified body of knowledge."

– *Biology*, a college textbook[1]

"The success of the natural sciences have led one analyst after another to extend Darwinian thinking into a series of distinctly nonbiological enterprises, even into the study of religion."

– Kenneth Miller, *Finding Darwin's God*[2]

As serious as it is to even begin pondering how to prepare to meet one's maker, expect no help from the atheistic sages of the twenty-first century. As far as they're concerned, there's very little reason to believe there's a maker to meet.

And this pervasive agnosticism, as we've already established, runs deep within modern scholarship. True, some intellectuals like Gould seem reluctant to embrace completely an outlook that denies the possibility of a wondrous Creator and his ministering angels. On the other hand, unbelievers such as Dawkins and his ilk seem barely unable to disguise their glee in brushing aside religion in favor of a secular faith.

What the Goulds and Dawkinses of science hold in common, however, is their firm belief that materialistic methods of inquiry have thoroughly eclipsed religious explanations of reality — however ancient and well-founded. In Darwinism, especially, they claim to have found the key to a fundamental understanding of the very nature of our cosmos and the creatures in it. By championing Darwin's central ideas — random mutation and natural selection — many modern scientists confidently extrapolate evolution far beyond mere biology.

As Dawkins puts it: "Natural selection not only explains the whole of life; it also raises our consciousness to the power of science to explain how organized complexity can emerge from simple beginnings without any deliberate guidance."[3]

This is indeed heady stuff. Order arising from disorder, life from nonlife. Haphazard change yielding complexity, consciousness and ultimately an unquenchable yearning for knowledge.

Heady, yes. But is it sound?

Key Concepts

The new atheists insist Darwinism leaves no room for God. Two prominent evolutionary scientists disagree:

❖ Francis Collins says moral law and the design exhibited by the cosmos could only have come from God.

❖ Kenneth Miller sees the indeterminate nature of quantum physics as God's guarantee of free will.

Resistance

An increasingly vocal minority of scientists and scholars are making it known that though they respect the methodology and the findings of modern science, they repudiate the notion that studies of the physical realm have even begun to explain all there is to know about ourselves and the cosmos. Assuming the role of what Neil Postman terms "loving resistance fighters," these critics in particular refuse to characterize human beings as purposeless machines or the universe as a lucky accident.[4]

A few scientists, upon pondering the limits of their craft, have dared to proclaim their theistic faith despite working in what they confess is a hostile environment. As biology professor and textbook author Kenneth Miller explains: "The conventions of academic life, almost universally, revolve around the assumption that religious belief is something that people grow out of as they become educated."[5]

He adds:

> "From time to time I have to struggle to explain to my students, and even my colleagues, not only why Darwinian evolution does not preclude the existence of God, but how remarkably consistent evolution is with religion, even with the most traditional of Western religions."[6]

Others go much further in their rebuke of modern science. But first, it's instructive to consider in further detail just what Darwinian biology claims to have discovered about life — and God.

What Hath Darwin Wrought

Darwin's theory is really quite simple; in many ways it only seems to recapitulate the obvious. Life forms change over time, that is to say, from generation to generation. Offspring, in their physical and mental characteristics, differ slightly from their parents and their siblings. Those members of a particular generation who manifest favorable traits, which grant them some advantage for both survival and procreation, will likely pass those traits on to the subsequent generation. Given enough time, Darwin mused, these advantageous mutations could amount to significant change in organisms — even to the point of creating new species.

Darwin developed this idea when considering how humans obtained impressive changes when breeding animals such as pigeons or dogs for specific physical characteristics. During his cruise on the research ship *Beagle*, Darwin also noticed how some island species exhibited enough similarities with mainland species to suggest they had descended from them, but had subsequently altered into new, original forms. Probably the most famous example of island fauna displaying novel forms are the Galapagos finches which now bear Darwin's name. Scientists have since classified Darwin's finches into about a dozen distinct kinds, usually referred to as species.[7]

That there should be any question of whether or not this collection of remarkable birds on a remote Pacific archipelago constitute different species hints at just one of the many embarrassing trade secrets within Darwinian science. In this case, what appears at first to be a textbook example of Darwin's principles in action is tainted by the fact that scientists don't necessarily agree on what defines a species, or even why there should be such a classification.[8]

Nevertheless, bolstered by what appeared to be incontrovertible evidence of the power of random mutation and natural selection, Darwin dared to extrapolate. If nature were left to work its incremental alterations, he mused, why couldn't new and more complicated life forms spontaneously arise through the simple law of descent with modification? Indeed, why couldn't all life forms arise from a single, original progenitor?

Then, of course, followed the ultimate extrapolation. If natural processes could account for all life forms, and possibly even the origin of life, why even postulate a creator?

However audacious, Darwin's original theory lacked a key element. It was all well and good to imagine life forms simply changing in a beneficial direction, for example, as in the giraffe's neck lengthening over several generations. But as scientists — and even farmers who study the effects of stock breeding — could point out, the most basic principle of heredity is that organisms tend to produce similar organisms. And even when offspring differ from their parents, the changes tend to be minor and are almost always simply a variation of a clearly identifiable trait in some ancestor. What's more, the range of variation conforms to strict limits. A donkey interbred with a horse produces a mule, but almost always that mule is sterile. A tabby cat may give birth to a calico, but never to a jaguar.

For Darwin's theory to account for the amazing range of biological variation apparent in nature, then, it needed a mechanism that not only could generate novel characteristics, but could provide a way to preserve these characteristics so that they could be built upon by subsequent generations. That mechanism apparently was provided in the rise of genetics, a scientific field built on the rediscovery of Gregor Mendel's seminal work in a nineteenth-century Austrian monastery.

By the mid-twentieth century, principles of genetics were melded with Darwin's theory of evolution to produce what was christened the "modern synthesis." The new hypothesis declared, as every biology student can today affirm, that evolution functions by first creating mutations on the genetic level, thus producing the so-called "raw material" for natural selection to play with and eventually build mice from microbes, elephants from eels.

With further discoveries into the structure of DNA and the inner workings of the individual cell, the triumph of Darwinism seemed complete. Indeed, DNA established the presence of a nearly universal genetic code that apparently needed only to be tinkered with a few bits at a time to lay the blueprints for just about every life form that has ever existed on Earth. In addition, the appearance of similar cellular structures and mechanisms across a wide range of organisms seemed to confirm the interrelationship of all living things.

Now that secular Darwinism had seemingly conquered physical biology, it only seemed fitting for some thinkers to apply its principles to other fields. Some like Ernst Mayr, one of the founders of the modern synthesis, insisted it was only logical to consider human morality as simply another product of unguided evolution.[9] Dawkins, astounded by the vast amount of DNA that apparently codes for nothing useful, famously wondered if the instructions for life were just the byproduct of "selfish genes" locked in a Darwinian struggle for their own survival. Building on this naturalistic assumption, he and other ultra-Darwinists have gone on to cast doubt on such fundamental suppositions about human nature as consciousness and free will. Indeed, if our bodies are merely the product of chance and necessity, why shouldn't our minds, thoughts and choices be viewed the same way?

Except that the same advances in biology have also illuminated inadequacies within Darwinism. Genetics and molecular biology, for instance, now reveal that the coding for life, as well as its most fundamental mechanisms, are more complex than ever could have been imagined. And this complexity is such that it defies not only common sense to believe it could have arisen by mere chance, it also trumps the efforts of mathematicians to explain it in terms of probability. Stymied by the intricacy of DNA in even the simplest life forms, for example, Darwinists have been compelled to search for simpler alternatives as the supposed genesis for genetic programming in all of life. Similarly, instead of postulating how the first complete creature was formed (the usual suggestion is a single-celled microbe), microbiologists are turning their focus toward describing how individual, highly regulated cell functions first evolved. What finally brought these divergent, but necessary, components together into a living, self-replicating whole is a tale, as we shall see, that Darwinists aren't completely comfortable committing to definitively.

> **Key Quote**
>
> The complexity uncovered within DNA is such that it defies not only common sense to believe it could have arisen by mere chance, it also trumps the efforts of mathematicians to explain it in terms of probability.

What's more, these and discoveries in other scientific fields have supported a wider challenge to materialistic orthodoxy. Findings in cosmology, mathematics, and even in the esoteric field of quantum physics have lent new hope to those who question the assertion that humans are just higher animals on a mundane planet in a universe governed by indifference. For these informed believers, science doesn't threaten faith — it bolsters it.

Can't Take That Away From Me

As religion scholar Ian G. Barbour explains: "A number of theologians and scientists have used concepts from recent science to portray God's relation to the world. God can be viewed as the designer of a self-organizing process or a communicator of information."[10]

Two scientists in particular confirm Barbour's general observation, proclaiming in a pair of recent books both their commitment to Darwinism and to belief in God. One of these is the aforementioned Miller, whose adherence to evolutionary biology is so keen that he was called upon as an expert witness in the 2005 federal lawsuit brought to prevent the theory of intelligent design from being taught in Dover, Pennsylvania public schools. The other is Francis Collins, who led U.S. government research efforts in the project to map the human genome.

Admittedly, Collins and Miller draw heavily on philosophy in defending their theistic faith. Yet both feel strongly that findings from within science — though in some cases, negative findings — make the supposition that God exists and is concerned with the fate of his creation a very reasonable one.

For Collins, the most powerful evidence for belief in God comes from outside his own area of expertise. (Biology, particularly genetics, Collins sees as thoroughly confirming evolution. He offers as proof various examples, such as the fact that certain mammals contain strikingly similar genetic material in their DNA).[11] From the realms of cosmology and physics, Collins points out findings which we shall discuss later in more detail, namely, the big bang theory and the so-called anthropic principle.

In brief, the big bang theory draws on observations such as the expansion rate of the universe, as well as the uniform presence of faint microwave radiation, to conclude that at the beginning of time the physical universe essentially burst into existence from nothing. The anthropic principle declares that the various fundamental forces which define how the universe functions have been fine-tuned with incredible precision for no apparent reason other than to support life. Though many of his colleagues distort the anthropic principle into a logical fallacy upholding materialism, Collins sees it as confirmation of a divine hand at work.

He writes:

> "If one is willing to accept the argument that the Big Bang requires a Creator, then it is not a long leap to suggest that the Creator might have established the parameters (physical constants, physical laws and so on) in order to accomplish a particular goal."[12]

And for Collins, the ultimate clue illuminating that divine goal is something science can never discount — moral law. In fact, Collins says it was the obvious truth of man's duty to disdain evil and strive for good that helped persuade him to embrace the doctrine of a Christian God who both conquers sin and enables righteousness. The added doctrine that God accomplishes this redemption through selfless love — and that humans are called to emulate God's example through selfless acts of their own — Collins says, confounds science.

> **Key Quote**
>
> Collins says it was the obvious truth of man's duty to disdain evil and strive for good that helped persuade him to embrace the doctrine of a Christian God who both conquers sin and enables righteousness.

"Agape, or selfless altruism, presents a major challenge for the evolutionist," Collins observes. "It is quite frankly a scandal to reductionist reasoning. It cannot be accounted for by the drive of individual genes to perpetuate themselves."[13]

Miller, as well, sees in the power of humans to choose to do good or evil powerful evidence for God's existence — evidence that scientific determinism cannot blot out. Still, Miller often finds himself facing a severe rhetorical dilemma in attempting to defend orthodox science while simultaneously delineating the limits of what it can know. On one hand, Miller insists, materialistic science has triumphed in explicating, through Darwinian evolution, how humans and other life forms arose without the aid of direct, divine creation. Yet Miller decries efforts by some scientists to explain human morality and behavior as irrelevant manifestations of the same unguided process.

For Miller, the key to understanding how God permeates the universe he made — and grants free will — is in the near-mystical realm of quantum physics. Here, where even the densest materials are mostly space and subatomic particles are not necessarily bound to behave in a logical way, Miller finds room for the Creator to work in a manner that need not interfere with science.

"The indeterminate nature of quantum behavior," Miller writes, "means that the details of the future are not strictly determined by present reality. God's universe is not locked into a deterministic future, and neither are we." He concludes: "In the final analysis, absolute materialism does not triumph because it cannot fully explain the nature of reality."[14]

Keeping Hope Alive

That Collins and Miller can cling to faith while backing evolution — and enduring the antagonism of their colleagues — should offer hope to non-scientist believers. And yet, for the faithful who wish to remain so while still granting scientific inquiry its due, the good news is that there is still more reason to hope.

As we shall see, the resistance to the materialist dogma of evolution runs much deeper than the ideas of either Collins or Miller. Indeed, upon reviewing the evidence, a good number of scientists and other thinkers are not only rejecting the philosophical conclusions of Darwinism, but are questioning the very premise of the biological theory itself. And this deeper resistance is not necessarily founded on any particular religious creation story — though certainly, a respect for religious truth does play a factor. This rising rejection is based on the fact that a comprehensive survey of nature makes it seem more and more as if it came into being through intentional, intricate design.

Chapter Four

Gaps in the Record

How Evolution Leaves Scientific Mysteries Unexplained

"The evidence, so far at least and the laws of Nature aside, does not require a Designer."
— Carl Sagan, *Pale Blue Dot*[1]

"Surely it is rational for people who believe that God is or may be the creator to challenge those who insist that we assume that a mindless nature did all the creating."
— Phillip E. Johnson[2]

In 2004, eminent British philosopher Antony Flew made international headlines by announcing a conversion of sorts. A longtime atheist, Flew revealed that, after considering the arguments of a particular group of dissenters from orthodox science along with his own observations, he no longer considered materialism an adequate explanation for the existence of the universe. Flew declared he now believes it likely that an intelligent force had a hand in forming the cosmos and the life in it.[3]

For proponents of intelligent design — the aforementioned dissenters whose articulate criticism of Darwinism have made their movement a household word — Flew's defection certainly presented a public relations coup. Just how profound Flew's change of heart is to him on a personal level, however, remains to be seen. He has been reluctant to identify himself as a believer in a sovereign deity or in life after death, preferring instead to characterize himself as a deist in the vein of Thomas Jefferson.[4] As such, it is safe to say that Flew is still a long way from being transformed into a religious apologist.

Still, the truth that Flew's tentative conversion was not fueled by a religious epiphany in one way makes it an even bigger score for critics of Darwinism. After all, atheists such as Dawkins have long relied on characterizing their opponents as ignorant fundamentalists who either can't or won't embrace the revelations of materialistic science. That an intellectual such as Flew is willing to challenge materialism based on the same system of deduction that once persuaded him to accept it as reasonable, lends great credence to the charge that the issue is one of fact.

> **Key Concepts**
>
> When philosopher Antony Flew announced his conversion from atheism to belief in an intelligent designer, it spotlighted problems with origin of life theories now favored by materialists. These include:
>
> ❖ Natural forces alone cannot explain how life emerged from nonlife or how complex genetic information came to be.
>
> ❖ Contrary to the predictions of Darwinism, the fossil record does not show the gradual, continuous formation of new species.

Much of what Flew cites as the collective impetus for his conversion we will consider in depth later. For now, a summary of his current thinking in regard to theism serves as a sort of index of future attractions. This is especially true in respect to Flew's own assertion that material observations — the evidence apparent in nature — ultimately composed just a part of the argument that persuaded him to change his mind.

Still, says Flew, science and its study of nature did provide a compelling starting point. To begin with, human observation of the cosmos raises the question of why nature — and human beings — even exist. Compounding this enigma is the fact that nature obeys laws. Flew writes, "The important point is not merely that there are regularities in nature, but that these regularities are mathematically precise, universal and 'tied together'."[5] Moreover, these laws function in a way that produces life, and not just any kind of life, but "intelligently organized and purpose-driven beings."[6]

Flew goes on to say that his recognition of such exquisite design in nature made him sympathize with modern scientists who attribute the order they observe to a higher power, a creative force often characterized as the Mind of God. And though Flew insists his journey toward belief in this higher power was indeed "a pilgrimage of reason," he confesses that belief in God cannot be attained by logic alone. He declares: "Those scientists who point to the Mind of God do not merely advance a series of arguments or a process of syllogistic reasoning. Rather, they propound a vision of reality that emerges from the conceptual heart of modern science and imposes itself on the rational mind. It is a vision that I find compelling and irrefutable."[7]

In other words, the contention that God does not exist, and that humans are a lucky accident, might be based on materialist theories including Darwinian evolution, but more and more it looks like these theories have little to do with the evidence.

Origin of Life

Darwinian evolution unravels from the very beginning, in failing to offer a naturalistic explanation for the origins of life. A cursory look at the study of life origins exposes how Darwinism as a whole relies a great deal upon extrapolation and intellectual sleight of hand.

True, there is a scientific orthodoxy of sorts when it comes to answering the fundamental question of how life on Earth began. But even this orthodoxy amounts chiefly to speculation. And when some scientists are pinned down on the question concerning what they actually know about how life emerged, they sometimes prevaricate and say that the topic doesn't really fall under the purview of evolutionary biology anyway. Or they fall back on the Darwinian adage (co-opted from modern geology) that since the present is the key to the past, and since we know that evolution has occurred, there necessarily must be a naturalistic explanation for life origins. This kind of circular reasoning appears frequently in Darwinian apologetics.

Darwin himself — who knew nothing of DNA or the requirement that every living cell be provided with genetic instructions — imagined some protean microbe coming to life in "a warm little pond."[8] Darwin's iconic image lives on in the modern myth of the chemical-rich prebiotic soup that somehow provided the raw materials for life's genesis — supposedly some 3.8 billion years ago.[9]

Advances in molecular biology, however, mean that the modern notion for what had to occur before even the simplest single-celled organism could be animated is exquisitely more complex than Darwin could have supposed. At the very least we have to imagine proteins and other large molecules assembling to form the cell's physical structure, as well as means for acquiring and metabolizing energy, and reproducing. All of the cell's structure and functions, though, first would have to be programmed in its DNA — a problem that foreshadows a common paradox within evolutionary theory. Indeed, in order to circumvent the obvious difficulty of trying to explain life without DNA, some scientists have proposed a fantastic world of organisms initially programmed not by the now-familiar double-helix, but a genetic code based on the simpler nucleic acid that serves as DNA's molecular scribe and messenger — RNA. How RNA evolved, however, is anyone's guess.[10]

Meanwhile, the best scientists can do at the moment toward a theory of life origins is to refer to experiments of the sort first conducted in the 1950s. Then, Stanley Miller and Harold Urey rigged an apparatus that was supposed to simulate atmospheric conditions on Earth before the emergence of life. The duo zapped a mixture with electricity to simulate lightning, and discovered that the process had produced, among other things, several amino acids — the building blocks of protein. Scientists today question whether the gaseous mixture Miller and Urey used truly mimicked the atmosphere on early Earth. Nevertheless, similar subsequent experiments apparently have produced the entire range of amino acids, plus several sugars, lipids and nucleotide bases for DNA and RNA.[11]

But it's a long way from complex molecules to living creatures. As one trio of college textbook authors put it: "Although biologists generally accept the hypothesis that life developed from nonliving matter, exactly how this process, called chemical evolution, occurred, is not certain."[12]

And for good reason. As Stephen C. Meyer points out, advances in mathematical theories of probability show that even the simplest molecules needed for life couldn't have arisen through pure chance. Simply put, there hasn't been enough time since the beginning of time for random forces to have run through all the possible combinations.

Meyer writes:

> "Significantly, the improbability of assembling and sequencing even a short functional protein approaches this universal probability bound — the point at which appeals to chance become absurd given the 'probabilistic resources' of the entire universe. Further, making the same kind of calculations for even moderately longer proteins pushes these measures of improbability well beyond the limit."[13]

Key Quote

Advances in mathematical theories of probability show that even the simplest molecules needed for life couldn't have arisen through pure chance. Simply put, there hasn't been enough time since the beginning of time for random forces to have run through all the possible combinations.

The daunting reality of life's improbability, however, does not prevent scientists from speculating about how some self-organizing principle must have arranged the chemical parts into the first living cell. Or suggesting that under certain conditions life may just be inevitable. Some theorists who despair of finding an explanation on Earth go so far as to look to outer space as a potential source of life. Nobel Laureate Francis Crick, who co-discovered the structure of DNA, for a time touted the concept of panspermia, the notion that aliens seeded life in likely places across the cosmos.[14]

For most of us, though, relying on super-advanced space aliens as the progenitors of life probably smacks too much of a meddling Olympian pantheon to seem like science. Still, the idea that terrestrial biology had its beginnings elsewhere must hold genuine appeal, because it has been advanced in various forms a number of times. In 1996, for instance, a team of scientists caused quite a stir by announcing what they thought was evidence of ancient life being transferred to Earth from Mars. The theory was based on a presumed meteorite discovered in Antarctica in the 1980s. The rock contained what could be interpreted as the fossil remains of a microbe — not necessarily the microbe that launched evolution, but perhaps evidence enough that life here could have come from out there.[15]

Except, of course, that scientists had to imagine how such a microbe arrived on Earth to begin with. In the case of the Antarctica meteorite, presumably the rock and its microbe were blasted from Mars by the impact of some other meteorite, then survived both the transit through the extreme cold of space and the fiery descent through Earth's atmosphere in order to possibly continue evolving on this planet. This presumption unraveled, however, when upon further scrutiny critics argued that there was no way to be sure the marks on the celebrated meteorite were made by anything that was ever alive.[16]

No Easy Explanation

Returning to the more orthodox "chemical evolution" theory of life's origins, however, in no way reduces the many difficulties confronted by Darwinian apologists. If indeed, the scenario suggested by Miller-type experiments actually occurred, it would had to have taken place in what scientists call a reducing (non-oxygen) atmosphere. Oxygen can be both volatile and corrosive, and its free presence in the atmosphere would have worked against the formation of the large organic molecules needed in building living cells. This means the first microbes would have had to be anaerobic — able to metabolize and otherwise function in the absence of oxygen. (Oxygen is usually toxic to such creatures.) Also, without the presence of any other organic matter, they would have had to be able to feed off the chemicals they presumably evolved from. (Either characteristic — being anaerobic or a chemical feeder — is fairly rare in current life forms.) Presuming that the microbes multiplied faster than the chemicals, these creatures then would have had to evolve into organisms that transform sunlight into energy and give off oxygen as waste. This arrangement could have helped transform the atmosphere into the oxygen-rich environment that supports life as we know it today.[17]

The obvious problem with such a view is that it relies on a succession of truly mind-boggling contingencies. The formation of organic chemicals by some unknown process. The presence of a non-oxygen atmosphere conducive to this unknown process. The encryption of genetic information by an unknown process. The rapid evolution of the first life form that was unlikely to survive unless it quickly adapted into something else.

These difficulties alone should be enough to prompt a reasonable person to consider the origin of life by purely naturalistic means all but impossible. And yet there is one more contingency to be considered that should expose Darwinian theories of biogenesis as utterly, irrevocably improbable — the fact that no one can suggest how life might spontaneously emerge from nonlife.

As popular science writer William Hartmann put it, when considering life origins:

> "The greatest current enigma is the 'magic' step between those nonliving, cell-like systems and truly living, reproducing cells.
>
> I say 'magic' not because we believe that the transition was supernatural (whatever that means) but merely because we still don't understand it."[18]

On the contrary, Hartmann's choice of words in summarizing what little science actually knows about where life came from is quite telling. Indeed, there is nothing within human knowledge that can explain life emerging from nonlife. Consequently, to suggest that it ever did is to invoke the supernatural.

Fossil Record Problems

Despite difficulties in accounting for life origins, Darwinists remain confident that the evidence supports their theory that living things on Earth gradually evolved from simple to more complex forms over vast amounts of time. The record of this process, they insist, is imprinted in stone. All one has to do is go dig for it.

And when examined in a cursory way, the fossil record does seem to offer strong support for Darwin's theory. Earth's oldest geological layers yield the remains of microbes and bizarre but relatively simple sea creatures. Middle layers yield bony fish, insects, amphibians, early plants and, of course, those terrible lizards — the dinosaurs. Only in the latest layers do birds and mammals, including humans, finally appear.[19]

Given that two branches of science — geology and biology — seem to dovetail so nicely in support of evolution, why is it then that some Darwinists admit to finding the fossil record a problem that needs to be explained away?

The answer lies in understanding one of the central tenets of Darwinism — that evolution occurs gradually. Just how gradually, however, is apparently open to interpretation. Darwin himself spoke of "infinitesimally small inherited modifications,"[20] and Dawkins famously discusses major improvements such as eyes and wings developing one percent at a time. The only caveat that seems beyond debate in determining the rate at which evolution occurs is that no change from generation to generation can be so great as to appear miraculous.[21]

In the long term, of course, evolution is supposed to bring about astonishing changes in life forms. But the process must be seen as building minutely upon what has come before, through descent with modification. Fish fins might evolve into legs, for instance. Forearms might evolve into wings or even back into whale fins. However, if evolution is gradual and continuous, there should be a record of transitional life forms imprinted in the fossil record in the proper times and places.

What one would not expect to find in the fossil record are wholly new and complete life forms appearing without any apparent ancestors. Nor should particular life forms exhibit long-lived stability, only to suddenly disappear without spawning a transitional organism that could evolve into more complex forms.

> **Key Quote**
>
> The pattern that fossils exhibit is not continual, gradual development from organism to organism, but the sudden emergence of new life forms and the rapid extinction of others.

But this is precisely what a closer look at the fossil record reveals. True, there are a few remarkable fossils such as *archaeopteryx* that could be seen as transitional forms between major types. But these are rare, willfully interpreted, and usually controversial. In fact, the pattern the fossils exhibit is not continual, gradual development from organism to organism, but the sudden emergence of new life forms and the rapid extinction of others.

This lack of transitional fossils troubled Darwin himself, who nonetheless held out hope that science would eventually uncover what have come to be caricatured as "missing links."[22] But the problem of very real gaps in the fossil record continues to goad Darwinian apologists.

Gould, for instance, argued in 1994 that the fossil record in no way supports the human conceit that evolution is predisposed toward producing complex creatures like ourselves. Quite the opposite, Gould insisted the fossils show that, "every once in a while, a more complex creature evolves and extends the range of life's diversity in the only available direction. …But the additions are rare and episodic. They do not even constitute an evolutionary series but form a motley sequence of distantly related taxa."[23]

The major reason for this pessimism concerning fossils is the so-called "explosion of life" recorded in the Cambrian period. Far from indicating gradual development, this epoch unveils every major animal phyla, as well as some bizarre extinct forms, appearing as if from nowhere.[24]

As Gould puts it,

> "...all major stages in organizing animal life's multicellular architecture then occurred in a short period beginning less than 600 million years ago and ending by about 540 million years ago — and the steps within this sequence are also discontinuous and episodic, not gradually accumulative."[25]

Confronted with the fossil record's lack of conformity to the theory they want so desperately to be true, a few more fair-minded scientists have suggested modifying Darwinism to make it fall in line with the actual evidence. Biologist Richard Goldschmidt, for instance, scandalized his colleagues by suggesting that evolution may not work in a steady, incremental way after all, but through quick, substantial alterations. According to science writer Francis Hitching:

> "He proposed, and thus aroused the ire of orthodox evolutionists, that these changes had happened suddenly, through 'monstrous' mutations — the kind that produced fairground exhibits like two-headed sheep or stunted rabbits. He agreed that almost all of these would fail to survive, but just occasionally, a monstrosity would make the grade — and in this way a new species would emerge."[26]

Gould, as well, teamed with fellow scientist Niles Eldredge in the 1970s to propose a less radical modification to Darwinism dubbed "punctuated equilibrium." This view contends that most of the time life forms remain stable, and do not evolve. These long stretches of stasis are interrupted by "quick and quirky" episodes during which new species develop relatively rapidly. Just how rapidly remains uncertain, as Gould and Eldredge maintain they are speaking in terms of geological time, with epochs that can last for millions of years. Nevertheless, according to this model evolution occurs at least rapidly enough during any given episode of change to create few transitional life forms, which consequently are rarely fossilized.[27]

Thus, punctuated equilibrium seeks to prove a negative, cleverly turning what hasn't been discovered into evidence for evolution. Still, orthodox Darwinists remain divided over its merits, possibly because, as Phillip Johnson puts it, the theory seems to hint at ideas anathema to true believers in gradualism.[28]

But if Darwinists fear for the future of their orthodoxy, they have plenty of reason to do so. We have already noted the problems with materialistic models of life origins. Add to this the non-Darwinian patterns found in the fossil record, and the theory that life developed via an undirected, purposeless process again falters. But it gets worse. When what little evidence paleontologists actually have managed to dig up in support of evolution is examined with the skepticism it merits, the evidence for Darwin's materialistic triumph wanes as thin and fine as bone dust.

Chapter Five

Of Apemen and Mutations

Human Origins Controversies Reveal Just How Little We Really Know

"A mistake in geology, a mistake in paleontology, and a mistake in dating. But you correct as you go. It all comes out in the wash."

– Donald C. Johanson, *Lucy: The Beginnings of Humankind*[1]

"More, the lesson is that insatiable public curiosity about man's origins on the one hand, and deeply held scientific prejudices together with fragmentary and ambiguous fossil evidence on the other, combine to give each generation the ancestor it wants, and each paleontologist the fossils he seeks."

– Francis Hitching, *The Neck of the Giraffe*[2]

In the 1890s, a Dutch doctor named Eugene Dubois went to the island of Java in what is now Indonesia, hoping to be the first to uncover evidence of an ape-like evolutionary ancestor to modern humans. What he found — a skullcap, a femur and two teeth from something presumably very old and not quite human — was not nearly as important as what he ultimately contributed to modern culture.[3] Because with these few bones and a great deal of imagination, Dubois breathed life into the myth of the slouching apeman, the hirsute, savage progenitor of modern humans whose inevitable rise to genius and self-awareness supposedly began when he first chose to leave the trees and walk upright. Having birthed a new mythology, Dubois' second great achievement was to lay the groundwork for a novel branch of human inquiry — one that blends the meticulous calculations of modern science with the vague mysticism of ancient oracular arts.

This area of study — in which scholars employ arcane knowledge to sift through old bones and concoct stories about our past, has come to be known as paleoanthropology. It is also one of the most contentious fields in modern science — and a microcosm of all the pitfalls inherent in applying a Darwinian view to the interpretation of fossils.

Not that any scientist who studies fossils has it easy. Field work is tedious, sometimes dangerous and often unrewarding. There is great pressure to produce results — evidence of some wondrous new creature that will astound colleagues, wow the public and persuade deep-pocketed benefactors to finance continued research. This is why, among the various specialties of fossil studies, human origins offer one of the best prospects for professional success. Human beings understandably are deeply interested in where they came from and why they exist, consequently, purported discoveries of hominid fossils — and their subsequent analysis — tend to garner the most publicity.

> **Key Concepts**
>
> The study of human origins provides a microcosm of all the pitfalls inherent in applying a Darwinian view to the interpretation of fossils. Consider:
>
> ❖ So-called apemen fossils are extremely rare and are usually fragments.
>
> ❖ Nearly always these fossils cannot be dated directly. Instead scientists must rely on clever inferences derived from dating nearby objects.
>
> ❖ Consequently, the field is subject to much interpretation, to the point of provoking open bickering among scientists.

On the other hand, for the very same reasons paleoanthropology also provokes a great deal of controversy. And this controversy has plagued the field from its genesis.

It was Darwin himself, of course, who suggested that modern apes and humans share a common ancestry. And by the close of the nineteenth century, bones discovered in Europe seemed to confirm his hypothesis. Various remains inspired names for these presumed proto-humans, usually based on the area where they were uncovered — Neanderthal Man in Germany, Cro-Magnon in France. Subsequent analysis of these finds, however, determined that they were not so different from modern humans and hardly constituted the primitive, ape-like creatures imagined by Darwinism.

Then came Dubois and his Java Man — dubbed according to the practice of the time. After finding a champion in the German biologist Ernst Haeckel, Dubois set about publicizing his discovery as a momentous transitional species. A life-sized model was built, which was displayed in museums throughout Europe.[4] The response Dubois sought, however, was not forthcoming.

In the words of renowned paleontologist Donald Johanson: "Instead of triumph, he experienced bitter controversy. For years his fossils were the subject of furious debate."[5] Critics questioned Dubois concerning the credibility of his methods, his evidence, and his hypothesis. In the end, writes Johanson, the embittered fossil hunter stashed the bones of Java Man "under the floorboards of his dining room and for thirty years refused to show it to scientists or to speak to them about it."[6]

The trials of Dubois, however, in no way derailed the search for more evidence of human origins. By the early twentieth century, more bone fragments had turned up in China and South Africa. The problem then became a sort of embarrassment of riches. Discoverers all seemed anxious to declare their particular fossils completely new species, without much consideration of how these proposed species may have been interrelated or whether they simply represented varieties of the same basic biological form.

To address these problems, new methods of classification and more rigorous standards for dating and analysis were established. Since then, dramatic discoveries have propelled paleoanthropology into the status of cutting-edge science. Impressive finds of more complete remains — such as Johanson's "Lucy" fossil found the 1960s and the "Turkana Boy" skeleton, uncovered by celebrated rival Richard Leakey in 1984, have done much to help researchers close in on a standard model of human evolution. The evidence, it seems, has made the apeman myth secure.

But as is often the case, this particular myth is not necessarily anchored in reality. Indeed, after more than a century of gathering hominid fossils, paleontologists find themselves facing familiar frustrations concerning gaps in the record and a dearth of transitional species. As for a standard model of human evolution, it remains as fragmented and as open to interpretation as the bones it is based on.

Key Quote

To begin with, no evidence has yet been found of the predicted ancestor between modern apes and humans.

To begin with, no evidence has yet been found of the predicted common ancestor between modern apes and humans. The closest scientists have come to identifying such a creature is with *Ardipithecus*, among the oldest purported hominid genera as well as the very latest in celebrated apemen. In a revelation that merited a special October 2009 issue of the journal *Science*, a team of paleontologists unveiled a skeleton assigned to the species *Ardipithecus ramidus* — a set of bones so sensational scientists agreed they, like Lucy's, deserved a non-academic but marketable nickname. Thus "Ardi" was introduced to the public as a 4.4-million-year-old hominid who was neither particularly human-like nor chimp-like, but somehow evidence enough to keep the study of human evolution in the headlines.

This in itself is remarkable given the provenance of the fragments pieced together to fashion Ardi. The 125 fossil bits that make up this particular specimen of *Ar. ramidus* were first discovered in Ethiopia late in 1994. Scientists also recovered at the excavation site an additional 110 remnants believed to be of the same species as Ardi. That it took 15 years to research and assemble these bones attests to the doggedness with which paleoanthropologists pursue their craft — and hints at the atrocious condition in which the remains were found. By all accounts, Ardi's skeleton — even by the standards of most purported hominid remains — was extremely crushed and distorted. Her skull apparently had been trampled by ancient hippos; her smashed pelvis was described by one researcher as resembling "Irish stew."[7] Some bits were so fragile they crumbled when touched. Researchers resorted to scanning the bones and piecing together the resulting digital images.

Finally, from Ardi's bones as well as thousands of additional plant and animal fossils, researchers decided they had enough on which to base some rather fanciful deductions. Ardi could walk upright; she had relatively small canine teeth and lived in a grassy woodland dotted with hackberry and fig plants. According to anthropologist C. Owen Lovejoy, this means Ardi's male counterparts probably disdained the violent mating rituals common in apes and instead decided to woo females with gifts of fruit. This in turn led to the monogamous pairings of apemen and apewomen, which ultimately resulted in cooperative human culture as we know it today.[8]

That is, if Ardi was truly a human ancestor. As *Time* magazine reporters covering the Ardi unveiling admitted, constructing the detailed social behavior of a long-extinct species based on fossil splinters is an exercise that at best can be described as "speculative." Anatomy professor Bernard Wood goes further, arguing that the evidence so far does not rule out the likelihood that Ardi represents a creature that evolved outside the line of humans.[9] Even Tim White, a paleoanthropologist and Ardi team member, confesses the possibility that Ardi was not even an evolutionary precursor to *Australopithicus.*[10] Before *Ar. ramidus* can be admitted into the Darwinian orthodoxy, it would seem the bones in question must first endure additional scrutiny, not a happy prospect given the fate of so many other would-be proto-humans.

For now, then, it is probably best to proceed to the next oldest hominid genus, *Australopithicus*, encompassing some six species (at least) over an estimated time range of 1 million to 4 million years ago. Scientists, however, disagree over which species of

Australopithicus — if any — can be considered a direct ancestor to humans.[11] At least one, ape expert Solly Zuckerman, questioned whether australopithecines could be considered human-like at all.[12]

Another critic, science writer William Fix, takes Zuckerman's doubts to an acerbic extreme. Ticking off the facts about *A. africanus* in particular — incongruous fossil dates, problems with anatomy — Fix satirically declares that the best that can be said about this celebrated species is that it might have been ancestral to modern chimps, but certainly not to humans.[13] In making this claim Fix inevitably pits the views of Leakey and his father against those of Johanson and his camp, exposing another common (and some might say, entertaining) problem frequently undermining paleoanthropology — insults, infighting and other manifestations of professional rivalry. Though much more could be said about this topic, for the present it's most apropos to address the confusing model for our own genus — *Homo.*

Again, a glance at the evolutionary charts in textbooks shows some half-a-dozen species of *Homo*, arising perhaps some 2.3 million years ago (and oddly overlapping with at least three australopithecine species). But in this case an initial glance is misleading. Three species, ours, Neanderthals, and so-called archaic *Homo sapiens* (sometimes labeled *Homo heidelbergensis*) are often considered more or less the same thing. *Homo erectus*, meanwhile, which may or may not be the direct ancestor of *H. sapiens* (since the archaic version of ourselves apparently overlapped with *H. erectus*, too) is sometimes split into another species, *Homo ergaster.* Finally, there's the toolmaking *Homo habilis*, usually presented as the oldest and founding member of the human family — except by those scientists who argue that *H. habilis* is really australophithicene.[14]

In summary, strip away the extraneous species, and from non-existent ape-human common ancestor to twenty-first century paleoanthropologist there are as few as five or six distinct evolutionary stages. Toss the extras back in and the total count of hominid species expands to a couple dozen, though no scientist can say for certain whether any one is the direct evolutionary progenitor of another.

This muddled modern view, believe it or not, represents an improvement over past depictions of human evolution. These historic difficulties Fix illustrates in the frontispiece of his book *The Bone Peddlers*, which charts a dozen proposed ancestors to man in order of their discovery, from Neanderthal to Lucy. All of them he declares problematic. Piltdown man, for example, was discarded as a phony apeman after being exposed as an outright hoax. Likewise *Hesperopithicus* suffered an ignominious unmasking when its bones were re-examined and found to be those of an extinct pig. Others, such as *zinjanthropus*, have been eclipsed by new discoveries.[15]

Key Quote

In light of these difficulties, it's easy to see why scientists today have rejected the quaint idea our lineage can be depicted in terms of a family tree. Darwinists today are more likely to invoke the image of a brushy shrub — multiple branches with few obvious connections that often terminate in evolutionary "dead ends."

In light of these difficulties, it's easy to see why scientists today have rejected the quaint idea our lineage can be depicted in terms of a family tree — with modern species branching off a few clearly defined, founding life forms. Instead, when offering metaphors for human origins, Darwinists today are more likely to invoke the image of a brushy shrub — multiple branches with few obvious connections that often terminate in evolutionary "dead ends."

Why the Confusion?

How did such an important field of science arrive at such a state? The reasons are varied and not always obvious. But for the most part the flaws within paleoanthropology simply echo the underlying problems within Darwinian science as a whole. Its adherents are so dedicated to a philosophical presupposition that when they do uncover potential evidence for their theory, they often are unable to tell when their analysis of that evidence passes beyond what is reasonable and lapses into wishful thinking.

As scientist Richard Lewontin summarizes: "No one knows or ever will know with the sort of evidence upon which we now depend, whether any fossil is a direct ancestor of the people who dig them up and write books about them. That will not stop the claims."[16]

Here are the facts. Hominid fossils — or even what can be reasonably described as hominid fossils — are extremely rare. One college textbook places the total number of sets in the hundreds.[17] Science writer Roger Lewin describes the world's inventory as "pitifully small."[18] Since most of these fossils are little more than bits and pieces, they require a great deal of imagination to analyze and reconstruct — which can be problematic.

As Johanson puts it, in contrast with his own discovery — the 40-percent complete Lucy skeleton, other fossil finds don't necessarily inspire confidence. He writes:

> "Yes, there are other hominid fossils, but they are all fragments. Everything that has been reconstructed from them had to be done by matching up those little pieces — a tooth here, a bit of jaw there, maybe a complete skull from somewhere else, plus a leg bone from some other place. The fitting together has been done by scientists who know those bones as well as I know my own hand. And yet, when you consider that such a reconstruction may consist of pieces from a couple of dozen individuals who may have lived hundreds of miles apart and may have been separated from each other by a hundred thousand years in time — well, when you look at the complete individual you've just put together you have to say to yourself, 'Just how real is he?' "[19]

Further troubling to those who would piece together man's past is that, almost without exception, the fossils they discover cannot be dated directly. And more than anything else, dating is key to Darwinian theories of human origins because scientists must be able to illustrate incremental progression from apes to us.

Without direct dating evidence, scientists are forced to rely on clever inferences drawn from other methods — mapping of geological strata, the proximity of other, more familiar fossils, and finally, radiometric dating. This last method is most favored, because it's based on what are considered extremely reliable calculations — the amount of time it takes for certain radioisotopes to decay. Again, the method does not date the actual fossil, but assuming the fossil is found near enough to a substance that can be subjected to radiometric dating, paleontologists feel fairly safe in applying the results to their own discoveries.

Unfortunately, it seems that even radiometric dating is more art than science, an unpalatable truth illustrated by the controversy that arose during efforts to date a fossil uncovered in 1972 by Richard Leakey. The skull in question, presumed to be a specimen of *Homo*, exhibited a larger braincase and more modern features than *Homo habilis* fossils found by Richard Leakey's father, Louis Leakey, also a famed paleoanthropologist. However, Richard Leakey's fossil originally dated at about 2.9 million years old — a full million years older than the *H. habilis* finds.[20] What's more, a date of 2.9 million years would have established the *Homo* line before most species of its presumed ancestors — australopithecines.

Luckily for those who did not wish to rewrite the history of human evolution, even the geophysicists who tested the volcanic ash where Richard Leakey's fossil was found — Cambridge University's Jack Miller and Frank Fitch — were dissatisfied with the results. The problem arose when subsequent tests by the duo yielded younger and younger dates. Lewin recounts, "they presented 41 separate age determinations … which varied between 223 million and 0.91 million years. Only seven of those 41 measurements came within a quarter of a million years either way to the original 2.61 number, while eight were as close to 1.95."[21]

Another independent radiometric testing of ash samples at Leaky's site, as well as the cross-referencing of animal fossils found there, finally yielded a consensus date for the hominid skull of about 1.6 million years. Still, one can't help but consider the dating fiasco and ask: What went wrong? Lewin cites one scientist who blamed Miller and Fitch's erratic results on their use of esoteric lab methods.[22] Johanson quotes another dating expert explaining that the original sample of volcanic ash yielded the wrong age because it "was contaminated by a few grains of much older material."[23] In other words, the samples must have been tainted because they failed to yield the anticipated results.

As for his own finds, Johanson reports that a second testing of basalt found near the Lucy fossils happily narrowed the range of possible dates and confirmed an age of around 3 million years — despite initial readings of 2.8 million to 3.2 million years old. In Lucy's case, the fear apparently was evidence of too young a date. In fact, the analyst rejected some volcanic ash samples because he worried they were probably "so badly weathered that they were useless."[24] In other words, it was decided ahead of time to avoid running tests on ash samples that probably would not support the range of dates the scientists wanted to find.

To be fair, it should be noted that the finagling of fossil dates in both these examples only resulted in an outcome that was pretty much the consensus to begin with. And scientists rely on more than just radiometric analysis in arguing for their particular interpretation of fossil ages. Still, examples like these can't help but cast doubt on the accuracy of dating methods. One doesn't have to be a skeptic to wonder if the process of determining which bits of bone evolved first amounts to much more than an educated guess.

Lewin concurs. The controversy over dating the Leakey skull, he declares, "illustrates not only that it is possible to be wrong in science, even with the apparently straightforward task of obtaining a single date for a single volcanic tuff; but also that typically there is a degree of uncertainty in science that is not often made public, because it is contrary to the mythology of what science is supposed to be like."[25]

But, scientists not only often fail to admit to the uncertainty inherent in their craft, they often refuse to recognize when they are in error. Fix, for example, takes to task Johanson and the writers of much of the professional literature since the 1970s for continuing to champion Lucy as a direct human ancestor even when evidence hinting otherwise should have induced them to be more cautious. Fix argues that the dates suggested for Richard Leakey's much more human-like skull discovery establishes that a species of *Homo* "was contemporary with the earliest-known, firmly dated *africanus*."[26] What's more, specialized anatomical features in Lucy — long arms and a ledge of bone on the skull known as a "simian shelf" — seem to indicate a creature much more ape-like than hominid hunters are likely willing to concede.

> **Key Quote**
>
> Scientists not only often fail to admit to the uncertainty inherent in their craft, they often refuse to recognize when they are in error.

Ultimately, Fix declares, despite Johanson's claims paleoanthropology is not self-correcting. Its adherents scrounge up some old bones, then through the power of narrative extrapolate what little is known about them into fantastic tales that are subsequently seized upon and embellished by attending scribes. Once thus entered into the mythology, only through revelation of the most scandalous truth can a particular apeman fantasy ever be eradicated. Fix writes:

> "This is quite a strange science, all in all. Too often there is not even a pretense of objectivity; not even a hint that there really is more than one possible explanation. Anyone questioning the credentials of a particular missing link is portrayed as having missed the boat — until something embarrassing happens to discredit the ancestor.

> Then the 'ancestor' is put to one side as quietly as possible, and the profession moves on to other fossils."[27]

The skeptics, though vindicated by the truth, are never clapped heartily on the shoulder by those who bought into the myth. Their faith remains unshaken, and the skeptic remains a traitor to the cause.

Hobbits

Indeed, for a myth-busting example of why Darwinists' interpretations of human origins still struggle for credibility, all one has to do is revisit the archipelago where supposed apeman bones were first disinterred.

In 2004, more than a century after Dubois, scientists digging on the Indonesian island of Flores made an astonishing discovery. In the sediment of a cave they found several tiny adult human skeletons, not much more than three feet tall. Among these only one skull, about the size of a grapefruit, was recovered. Associating these remains with stone tools and animal bones also found on Flores, and determining all dated to about 18,000 years ago, Australian members of the research team concluded they had discovered a separately evolved species.[28] This species was dubbed *Homo floresienses* and nicknamed "hobbit."

The reaction unfolded predictably. The story made international headlines. *National Geographic* magazine gave it the full treatment, with an artist's rendering of the hobbit as a cunning savage and glowing quotes from elated scientists. As journalist John Vidal put it, the hobbit fossils were "greeted as the most breathtaking and important discovery in 150 years, changing our understanding of late human evolutionary geography, biology and culture."[29]

Then came the rebuttal. By early 2005, Teuka Jacob, a top Indonesian paleoanthropologist who had cached the fossils in the university lab he heads, was casting doubt on the entire hobbit analysis. Jacob suggested that the tiny skeletons were the remains not of some strange species, but of human pygmies, very much like the sort who still live on Flores. The miniature skull he explained as the result of a developmental disease called microcephaly. Other scientists conducting subsequent research agreed.

Jacob went on to accuse the Australian members of the hobbit team of shoddy science and unethical behavior. He said they overlooked obvious humans features in the fossils and had slighted their Indonesian colleagues by excluding them from the process of announcing the separate-species theory.

"I don't think the Australians have the expertise," Vidal quoted Jacob as saying. "They were very narrow. They have tunnel vision and are not equipped in this area."[30]

Australian scientist Richard Roberts countered with allegations of his own. He suggested Jacob deliberately kept the hobbit bones locked in a vault in order to stifle research that might undermine a model of human evolution he and his colleagues back.

"All…are supporters of the multiregionalism evolutionary model," Robert charged, according to Vidal. "This discovery would destroy their theory." Opposing the hobbit separate-species hypothesis "suits their purposes very nicely." [31]

Some years later, the imbroglio over the Hobbit bones has yet to be settled, probably because, as Roberts pointed out, it exposes a greater rift between scientists concerning human evolution. Simply put, evolutionary biologists disagree over whether species of the *Homo* line evolved in Africa and then spread across the globe, or whether they evolved more or less simultaneously in various places. In other words, was there a clear line of evolutionary descent or a hodgepodge of populations whose differences never really prevented them from interbreeding?

And if this particular debate stirs academicians to great passions, there's a good reason. The crux of the human origins controversy goes to the heart of Darwinism itself. After more than a century and a half, Darwinists still struggle with the fact that instead of revealing constant, gradual transformation from species to species, the fossils instead echo a pattern witnessed in living creatures — species displaying variation within a controlled range — but no more.

Consider the criticism surrounding the hobbit hypothesis. The scientists who argue that the Flores bones are indeed *H. sapien* insist the hobbit team failed to account for the normal range of variation manifested in human beings. The critics go on to say that on top of normal traits there are normal abnormalities that must be accounted for — in this case, microcephaly.

The extraordinary thing about this argument is the underlying philosophy no one dares mention. Indeed, according to strict Darwinism, abnormalities (otherwise known as mutations) such as microcephaly should be taken as indications of evolution at work. And yet here is one highly publicized fossil study where some evolutionary scientists insist that in terms of declaring evidence of Darwinism at work, ordinary, run-of-the-mill mutations just don't count.

What, then, can we conclude about the state of human origin studies as a whole? Geneticist Steve Jones, who in no way questions evolution, sums it up this way: "Paleontologists still do not agree about where modern humans came from and where they went. The fossil record is so incomplete that a cynic might feel that the main lesson to be learned from it is that evolution usually takes place somewhere else."[32]

Jones no doubt meant for his comment to be understood as more satirical than dismissive. But whatever the tone, the statement exposes a profound lack of confidence in the evolutionary model of human origins — what is supposed to be one of evolution's strong points. Indeed, when a Darwinist confesses sympathy for cynics it can't bode well for Darwinism as a whole, and it certainly does little to bolster those who would extrapolate naturalistic evolution into a substitute for God. If Java men and hobbits can be scorned as meager grounds for building a materialistic creation myth, then God help the new unbelievers as they resort to an even flimsier aspect of their theory — the claim that natural forces are capable of creating new species.

Chapter Six

Life is Complex

How Evolution Overlooks Biology's Intricate Design

"It is clear that here on Earth we are dealing with a generalized *process* for optimizing biological species, a *process* that works all over the planet, on all continents and islands, and at all times."
– Richard Dawkins, *The God Delusion*[1]

"Each of the anatomical steps and structures that Darwin thought were so simple actually involves staggeringly complicated biochemical processes that cannot be papered over with rhetoric."
– Michael J. Behe, *Darwin's Black Box*[2]

Darwinists not only have a hard time producing fossil evidence that shows gradual evolution from simple to more complex life forms, they have an even harder time accounting for how such evolution would occur.

As noted earlier, according to neo-Darwinism the mechanism that drives evolution is supposed to be random genetic mutation. And this supposition seems sound enough at first, since mutations do occur, and sometimes are even passed on to offspring without deforming or killing them. Whether or not these mutations, or more to the point — series of mutations — can succeed in developing increased biological variety and complexity is another matter altogether.

For instance, when Darwinists speak of genetic mutations providing the "raw material" for evolutionary change, their meaning is not clear. They are not necessarily referring to the most common source of genetic error — either mistakes in transcription of relevant sections of DNA during the production of proteins, or in copying of DNA during cell replication.

Mutations of these type certainly do occur, but they are extremely minor and mostly useless.

What Darwinism requires is a random means of producing a change either in form or behavior that is significant enough to boost the species' evolutionary "fitness" — a novel benefit that can be "selected" — and which then can be passed on to subsequent generations. But in life forms that reproduce sexually, mutations occurring from normal cell replication or other DNA-driven functions not related to reproduction typically do not constitute inheritable traits. (Species that reproduce asexually present different problems for Darwinists, mostly too little opportunity for mutation during each reproductive cycle. As geneticist Jones points out, for instance, plants that reproduce this way suffer rapid genetic deterioration within a few generations.[3]) Consequently, Darwinists have shifted their search for a potential evolutionary mechanism to developmental genes — the so-called switches that help control embryonic development.[4] Unfortunately, there are serious problems with this theory, as well, but first it may be instructive to back up and examine cellular mutation.

Key Concepts

The new atheists make Darwinism the foundation for their secular outlook, but the theory still lacks a definite mechanism for driving evolution.
❖ Scientist Michael Behe illustrates how even on the cell level living organisms are "irreducibly complex."
❖ Field studies show that species can exhibit great variation, but only within certain limits.
❖ Some Darwinists admit the limitations of genetic mutations and are focusing instead on little-understood controls for embryonic development.

Small Cells, Big Problems

On the subject of whether mutations on the cellular level can explain the emergence of complex biological functions, perhaps few critics have stirred such ire as microbiologist Michael Behe. In his seminal book, *Darwin's Black Box*, Behe attacked evolution on the grounds that even microscopic cellular machinery displays a level of coherent, interdependent design that never could have arisen through incremental, unguided changes.

Behe described systems such as the bacterial flagellum and the blood-clotting mechanism in mammals as displaying what he calls "irreducible complexity." This means they require a convergence of structure, information and function in order to work properly. Disable one part of these systems, and they fail as a whole. Since even these relatively basic systems appear to have been engineered for their specific tasks, Behe argued, it is reasonable to assume that they are the work of an engineer.

Behe pressed this design inference even further in his next book, *The Edge of Evolution.* Focusing on the biological duel between the malaria parasite and its human host, Behe concluded that although microevolution can convey advantages in terms of sheer survival, it achieves these benefits not by constructing improved "fitness," but by disabling otherwise useful functions that an enemy life form might exploit.

In the case of malaria, a mutation that alters the shape of human red blood cells effectively counters the parasites when they attempt to proliferate within the bloodstream. The manifestation of this mutation, a disease called sickle cell, combats the parasites by essentially trapping them in the deformed blood cells, which are then destroyed, along with the invaders, in the spleen. Unfortunately, sickle cell disease can hardly be labeled an evolutionary advancement because it is only effective in saving lives when it is inherited from one parent. When inherited from both parents sickle cell is often fatal.

Behe argues that sickle cell undermines the Darwinian myth of beneficial mutations. This is because sickle cell "mutations are not in the process of joining to build a more complex, interactive biochemical system." He adds: "The chaotic interplay of genes is not constructive at all."[5]

But there's another aspect to the malaria conundrum, namely, why the parasite has failed to evolve a means of circumventing the sickle cell countermeasure in human beings. As Behe points out, malaria parasites are one of the most opportune organisms for exhibiting favorable evolution because they are relatively simple and multiple rapidly. (A single human being can host a trillion malaria organisms.)[6]

For instance, mutations in a pair of amino acids associated with the parasite's digestive vacuole helped it develop resistance to the medicine chloroquine.[7] However, in the case of a more recently developed drug, pyrimethamine, scientists were surprised at how malaria parasites were slow to develop resistance. Though the parasite apparently needs a mutation in only one amino acid to develop minor resistance to pyrimethamine, to increase its defense against the new drug it requires additional mutations, which Behe says also "seem to interfere with the normal work of the protein."[8] Behe deduces from this example that by degrading the intricate systems engineered into cellular machinery, genetic mutations ultimately work against the progress that Darwinists claim they are able to achieve. It is for this reason that malaria parasites, despite having more opportunities for mutation than more complex creatures, have failed to evolve enough to overcome the obstacle of sickle cell disease.

There is yet another way in which the human battle against malaria is telling for Darwinism. Applying mathematical principles of probability, Behe calculates the approximate rate for genetic mutations in malaria parasites and extrapolates it into a model for how frequently evolution occurs in general.

As groundwork for these calculations, Behe cites medical experiments intended to ascertain how quickly malaria developed resistance to certain drugs. "For instance," he notes, "in the case of atovaquone, a clinical study showed that about one in a trillion cells had spontaneous resistance."[9] This resistance developed due to a single amino acid mutation in the malaria parasite.

Now, if the organism had needed additional mutations in order to gain resistance, Behe argues, the science of probability declares that the odds of achieving such a beneficial change would increase exponentially with every alteration required. That malaria cells have attained such resistance to certain drugs, however, speaks not so much of their excellent luck, but to having math on their side due to their short life cycle and prolific reproduction. The fact is, he declares, malaria parasites ultimately have failed to evolve enough "fitness" to overcome obstacles such as sickle cell disease. This proves that evolution of any kind happens rarely and produces results too meager to account for even a three-point change in genetic coding, let alone the appearance of entire new life forms as imagined by Darwinists.

Apply this fact to humans, Behe adds, and the verdict against the power of Darwinism is undeniable:

> **Key Quote**
>
> Assuming Behe is correct, and random mutations do little on the cell level to provide the groundwork for evolution, how do Darwinists account for the development of complex systems such as wings and eyes? The answer is: Mostly through guesses and rhetoric.

> "The likelihood that *Homo sapiens* achieved any single mutation of the kind required for malaria to become resistant to chloroquine — not the easiest mutation, to be sure, but still only a shift of two amino acids — the likelihood that such a mutation could arise in the *entire* course of the human lineage in the past ten million years is miniscule — of the same order as, say, the likelihood of you personally winning the Powerball lottery by buying a single ticket [emphasis in original]."[10]

More Complexities

Assuming Behe is correct, and random mutations do very little on the cell level to provide the groundwork for evolution, how do Darwinists account for the development of complex systems such as wings and eyes? The answer is: Mostly through guesses and rhetoric.

Dawkins, for instance, describes various biological marvels as the logical culmination of successive improvement. Human vision, he argues, is built on simpler organs such as light-sensitive photocells and animal eyes. Flight arose from proto-wings that defied gravity just enough to help an organism break a fall or glide a little. Once this march of progress began, natural selection assured it would succeed.

Dawkins insists, "Half a wing is indeed not as good as a whole wing, but it is certainly better than no wing at all. Half a wing could save your life by easing your fall from a tree of a certain height. And 51 percent of a wing could save you if you fall from a slightly taller tree."[11]

Dawkins's argument, of course, makes a sort of rudimentary sense — if applied to a bizarre world of uniformly sized mutants who live in constant danger of falling from trees of a certain height. But in our own world of biological fact, Dawkins' account of how Darwinian evolution could create flight from nonflight remains conspicuously lacking. How, for instance, did evolution develop the feathers, lightweight bones, and peculiar muscle arrangement needed to propel birds into flight? What about the simultaneous metabolic changes needed to fuel such a demanding activity? Or the accompanying cognitive and behavioral alterations required for an avian lifestyle?

And what about the evidence of such evolutionary advances? As we noted earlier, the dearth of transitional fossils illustrating the evolution of flight — in birds, mammals, or otherwise — poses no challenge to those whose primary tool for championing their theory is intellectual sleight of hand. (One could argue, for instance, that 26 percent of a cheetah leg is better than 25 percent of a cheetah leg, but that would still leave what is now the world's fastest land animal a good 74 percent short of what if needed to do anything more than hobble around the African savannah in vain pursuit of its quarry.) Moreover, if it seems like there is no possible way a current biological system could have evolved without the aid of some wholly absent precursor, no problem. If the necessary precursor is no longer evident, that's because it's no longer needed. Dawkins argues:

> "There are many structures that are irreducible in the sense that they cannot survive the subtraction of any part, but which were built with the aid of scaffolding ... no longer visible. Once the structure is completed, the scaffolding can be removed safely and the structure remains in place. In evolution, too, the organ or structure you are looking at may have had scaffolding in an ancestor which has since been removed."[12]

So, just as in punctuated equilibrium, what doesn't exist is somehow turned into evidence for evolution. Again, how well does this sort of rhetoric hold up when compared with the facts? According to critic Wolf-Ekkehard Lönnig, not very well.

Lönnig takes to task several Darwinists — especially Dawkins — for failing to honestly confront the difficulties present in concocting an evolutionary explanation for one particularly puzzling species, the long-necked giraffe. To begin with, the idea that the giraffe gradually evolved from an ancestor resembling the modern okapi just isn't supported by the fossil record. This doesn't prevent Darwinists from suggesting that the giraffe — despite the two-to-three meter height difference — is more or less an okapi with a stretched neck.[13]

But as Lönnig states, "Only in the fantasy world of evolutionary theory are things as simple as that."[14] In fact, arguing in a manner that echoes Behe, Lönnig points out that the giraffe's long neck and great height require a complex system of unique features that defy a gradualistic explanation. Consider:

■ To accommodate pumping blood to its far-flung extremities, the giraffe has a massive heart and maintains extremely high blood pressure. In order to counteract this high blood pressure in its legs, the animal also has thick blood vessel walls and skin — not unlike an astronaut's pressure suit.

■ To prevent complications as blood drains from and rushes back into its head as it bends to drink and stands upright, the giraffe's neck includes an intricate system of blood vessels with special muscular walls, valves, and reservoirs.

■ The giraffe has a higher breathing rate to compensate for the amount of time it takes for air to travel down its windpipe.

■ The giraffe has a special muscular esophagus to propel the partially digested cud from its stomach back to its mouth.

Given the giraffe's complexity, Lönnig argues further that there should have been many, many intermediate species preceding the modern animal. How many? Well, if there were only a single species for each millimeter in height difference between okapis and giraffes, that means there must have been at least 200 intermediate species. However, if we extrapolate this simple calculation based on paleontologist George Gaylord Simpson's estimate that it took horse teeth about a million years to evolve one millimeter, the problem compounds itself.

Lönnig writes:

> "According to the theory of gradual evolution, at least 1,000 links are missing between the okapoid ancestor and *Giraffa*, conservatively estimated. If one applies Simpson's considerations to the growth rate of the 7 neck vertebrae, etc. — more literally, i.e. with numerous links per millimeter — one can even postulate 10,000 or more links."[15]

Of course, if the gap between okapis and giraffes requires tens of thousands of species in order to be bridged, Darwinists might be quick to claim this as a confirmation of their theory. It is anything but.

Why are we to assume that evolution would advance in the same direction, adding improvement upon improvement for thousands of generations? Not only does it defy belief, it defies evidence.

This Far, and no Farther

Recall again that the aim of incremental, progressive changes supposedly produced by Darwinian evolution is to develop new species. If we grant that individuals within a species do exhibit variation (sometimes to a great degree), then the crucial question becomes this: Have scientists ever observed enough accumulated changes to produce even one new species?

The answer is that scientists aren't sure.

Long-term studies of a handful of species certainly have recorded some remarkable instances of variation and natural selection at work — including some changes that have made the difference between life and death. Perhaps the most famous of these studies focused on Darwin's finches during several droughts over several decades on one of the Galapagos islands.

Scientists Peter and Rosemary Grant observed that as the island food supply dwindled during these dry spells, so did the number of birds with smaller beaks. In Darwinian terms, this meant the desperate conditions "favored" birds with beaks thick and deep enough to access the remaining food source — large, tough seeds. As a consequence, in subsequent generations large-beaked finches become more common. Here, it seemed, was a classic case of Darwinian evolution in action. But any Darwinists who hoped they were witnessing progressive mutations that would yield a new species were ultimately disappointed. When the rains returned to Daphne Island, the finch population again shifted back toward "smaller finches with average-sized beaks."[16]

As science writer Carl Zimmer summarizes: "The finches can change quickly, but it seems that they are swinging back and forth like a pendulum. After tracking 4,300 medium ground finches on Daphne Island between 1976 and 1993, the Grants have found no overall trend in beak size."[17]

In other words, any population shift in bird or beak size was temporary, due to a pre-existing genetic template that employs variation as survival tool. This fact now makes the idea that prolonged environmental change could somehow morph an organism beyond its programmed limits seem like a piece of hopeful naivete.

At the very least such a theory presents just another baffling evolutionary paradox. If, for instance, it is indeed environmental change that applies the "selective pressure" which forces a species to evolve, what happens if the environment changes faster than the species does? (Extinction, of course, but this doesn't get us what Darwin promised to deliver — new species.) On the other hand, if the environment remains stable for the eons it supposedly takes to produce a new species through incremental change, then what selective advantage could the current species possibly gain by evolving?

Key Quote

If the environment changes too quickly, presumed gradual evolution can't keep up and species go extinct. On the other hand, if the environment remains stable for the eons it supposedly takes to produce a new species, then what selective advantage could the current species possibly gain by evolving?

It's in the Genes

The problem is that the steady biological change supposedly driving organisms toward improved adaptation and complexity in actuality occurs only in the minds of Darwinists. In the world of living things, as in the case of Darwin's finches, physical changes do occur, but they tend to happen quickly, within specified limits. And just as quickly they can undo themselves.

Why do physical changes in organisms oscillate within a predetermined range instead of progressing toward novel forms? The answer represents yet another blow to neo-Darwinism, this time to the very thing the theory champions as providing the "raw material" for unguided evolution: genetics.

Genes, of course, are the precursor for any physical characteristic — and perhaps for a great many behavioral characteristics as well. So before any new, advantageous physical or behavioral change can appear in an organism, the plan for that characteristic must be written in the organism's genetic code — its genome.

Darwinists argue that the genome is in some ways like a word scramble. Mix the letters up enough times and you're sure to arrive at a combination that makes sense. Add to the phrases and sentences you already have and you get increased complexity.

Nothing could be further from the truth. On the contrary, studies of how genetic information changes within certain populations only confirms the conclusions derived by studying changes in physical characteristics. The frequency of certain genes certainly does change under certain circumstances, but in correlation to this, certain laws of genetics work to counteract any extreme alteration. The end result, more often than not, is stasis or even genetic degradation — certainly not the new and improved genome imagined by Darwinists.

Consider these genetic principles from basic biology, all of which are considered factors in shaping what physical and behavioral characteristics will predominate in certain populations.

The first principle is simply an extrapolation of probability theory. It states that populations will change merely because individuals do not procreate in random fashion, but usually choose mates based on some criteria. The reason for choosing a mate could be as basic as proximity — it's certainly easier to pair up with someone nearby than someone far away. Or sometimes individuals choose mates based on a particular characteristic. For instance, some fruit flies tend to choose sexual partners who are like them in respect to whether they are hairy or not so hairy. In both these instances, populations will tend to become less diverse; certain traits will diminish and individuals will become increasingly alike.

A similar thing happens in a phenomenon known as genetic drift. This refers to the sometimes drastic reduction of genetic diversity that occurs when small groups of a certain species become isolated. Darwinists like to tout genetic drift as a prime candidate for causing speciation, because it has been known to profoundly alter both genomes and the traits they manifest.

Isolation can produce amazing adaptability in a population, as in polar bears. It can produce oddities, such as the "Hapsburg chin" named for one of Europe's socially isolated (and inbred) ruling families. And it can prove severely detrimental, as it has by radically increasing the frequency of a recessive gene for dwarfism in Pennsylvania's Amish population.[18] But never does isolation produce new genetic information — it simply exaggerates the effects of traits already coded for.

And we have already seen how quickly these genetic exaggerations can be undone by what scientists call gene flow. This simply refers to the introduction of genes from individuals who previously were not — at least as procreative partners — part of a particular group. As the rapidly normalized population of Galapagos finches illustrates, the infusion of genetic diversity actually works against the kind of long-term successive change so sought by Darwinists. In other words, the normal mechanics of sexual reproduction and genetic change are not likely to produce new species.

In a Quandary

Which brings us back to the quandary over a mechanism for evolution. Random mutation, of course, remains the darling of orthodox Darwinists, because it at least has the virtue of being able to rewrite small pieces of the genome. But as we've already noted, and as more scientists are willing to admit — the power of random mutation to bring about positive changes, let alone dramatic positive changes, is extremely limited.

To cite the biology textbook authors again, "As an evolutionary force, mutation is usually negligible," though they are quick to add "it is important as the ultimate source of variation for evolution."[19]

But if random mutation is so limited, how does evolution work? According to the same textbook authors, it just may work through those special genes involved in embryonic development. In other words, evolution bypasses the logjam of minute mutations by tinkering with the control panel for forming the next generation:

> "Evolution is a conservative process, and natural selection builds on what has come before rather than starting from scratch. The evolution of new features often does not require the evolution of new developmental genes but instead depends on a modification in developmental genes that already exist."[20]

And so the argument for Darwinian evolution comes full circle. On one hand, we're told that evolution is creative, able to form wings and eyes, whales and wombats, mindless microbes and introspective humans. On the other hand, scientists admit that evolution is such a conservative process that it needs special "genetic toolkits" by which it can make a few tweaks and produce astonishing biological novelties.

Key Quote

On one hand, we're told that evolution is creative. On the other hand, scientists admit that evolution is such a conservative process it needs a special "genetics toolkit" by which it can make a few tweaks and produce astonishing biological novelties.

But how can we be sure that the new infatuation with developmental genes gets us any closer to a true evolutionary mechanism than earlier theories? Science at this point is only beginning to understand the intricacies of embryonic development. And what has been revealed points to unimagined complexity. Like protein construction within the cell, the genetic mechanisms controlling the growth of embryos into viable juvenile organisms represent marvels of micro-engineering. First, genetic information from the egg must be interpreted so that body plans can be laid out. Then, certain genes trigger other genes so that development can occur in a meticulously timed progression. Finally, to make the matter more complicated, there is evidence that genes and DNA alone do not control embryonic development, but that other, unknown factors may be involved.

This could account for what intelligent design advocate Jonathan Wells says is the fact that genetic alterations in the embryo never result in favorably mutated offspring:

> "Biologists have found that mutations in developmental genes often lead to death and deformity, but they never produce changes that benefit the organism. Furthermore, DNA mutations never alter the endpoint of embryonic development; they cannot even change the species."[21]

If ultimately Darwinists are hard-pressed to prove that genetic mutations — at whatever pace they occur — are unlikely candidates for creating new species, then the claim that their theory makes a divine Creator superfluous becomes dubious indeed. What's more, with scientists such as Behe claiming the micro-machinery of cells points to a designer, the evidence of biology actually turns against atheism. And as we'll see, when we lift our gaze from the wonders of the microscopic world to the grandeur of the cosmos and its underpinnings, the evidence for design stretches as wide as the heavens.

Chapter Seven

Cosmology as Creation

The Universe Looks Like It Was Made With Us in Mind

"There are billions of planets in the universe, and, however small the minority of evolutionary friendly planets may be, our planet necessarily, had to be one of them."
– Richard Dawkins, *The God Delusion*[1]

"The Big Bang cries out for a divine explanation. It forces the conclusion that nature had a defined beginning. I cannot see how nature could have created itself. Only a supernatural force that is outside of space and time could have done that."
– Francis S. Collins, *The Language of God*[2]

The universe we live in had a beginning. Being finite, it is also likely that at some point it will cease to be.

Intuitively most of us recognize this to be obviously true. But to modern scientists — especially the atheists among them — admitting that the cosmos both began and will end carries with it the unpalatable philosophical implication that a material explanation for the universe is not enough.

Academicians steeped in the materialistic approach to inquiry would be much more comfortable if they could prove that the cosmos is infinite and eternal, or at least the product of some blindly repetitive process. But since the evidence supports neither conclusion, cosmologists are often forced to resort to one of two philosophical canards. They propose fantastic physical theories of creation based on what amounts to mathematical fiction, or they withdraw into nihilism of the sort personified by scientist Carl

Sagan, who famously insisted that the physical universe is all there is and all there ever will be.[3]

But here's the problem: As our knowledge of the physical realm increases, the more these dual attitudes are exposed as presumptuous and naive. The more scientists learn about how intricate the cosmos is, from basic particles and forces to the movement of heavenly bodies, and how incredibly fine-tuned it is to support life — namely human life — the more difficult it becomes to dismiss our universe as merely a fortunate arrangement of energy and matter.

Put simply, a finite universe requires a first cause. And a fine-tuned universe hosting intelligent life certainly requires a living designer.

> **Key Concepts**
>
> The new atheists are troubled by developments in cosmology because they seem to undermine materialist explanations.
>
> ❖ Current standard models describe the universe as finite, with a definite beginning.
>
> ❖ The basic forces that shape the cosmos seem to be fine-tuned to support life.
>
> ❖ Some theistic scientists argue that the Earth is extraordinarily well-suited to allow for discovery.

Clockwork Universe

Science has not always modeled the cosmos as finite and changing. Even Isaac Newton, after establishing through his theory of gravity that all objects with mass attract one another, apparently saw no reason to believe this meant the stars and planets would ultimately collapse in on each other. Near the end of the seventeenth century Newton argued that if the universe contained an infinite number of stars equally spaced apart, then gravity would hold them in place forever.[4] His was essentially a clockwork view of the universe, where astral bodies had been hung into place and spun into eternal motion.

True, a few scientists did question apparent contradictions inherent in viewing the universe as infinite. As one argument went, if the night sky were indeed filled with an infinite number of stars (an oxymoron, since whatever may be numbered cannot be infinite), wouldn't it be completely illuminated and indistinguishable from daylight? For the most part, however, such objections were brushed aside. According to physicist Stephen Hawking, up through the nineteenth century the majority of scientists didn't bother theorizing on how the universe came into existence. He writes, "It was generally accepted that either the universe had existed forever in an unchanging state, or that it had been created at a finite time in the past more or less as we observe it today."[5]

Just a few decades into the next century this changed dramatically. First, Newtonian physics were supplanted by the rather non-intuitive observations contained in Einstein's theory of relativity. Einstein argued that time is not an independent characteristic, but that it is interterwoven with space, forming a malleable fabric (conveniently called "spacetime") that is warped by any object with mass. According to this model, as objects move through three-dimensional spacetime, they create channels which other objects follow, thus accounting for gravity and the principle of mutual attraction. In this way, excessively massive objects could generate a powerful enough gravity sink to warp anything — including the closest thing in the universe to pure energy — light itself.

To a genius like Einstein, the implications of his theory — in terms of how it affects cosmology — should have been obvious. If gravity was indeed produced by the very shape of space, then the notion of a static, clockwork cosmos could not stand. Without the momentum induced by some primeval force, which science could not account for, eventually the astral bodies must converge into a single mass.

But Einstein apparently considered this scenario too outlandish. In what has come to be known as one of history's most infamous acts of intellectual timidity, Einstein revised his model to comply with his preferred view of the cosmos. Hawking explains: "Rather than give up such an everlasting universe, which he and most other people believed in, he fudged the equations by adding a term called the cosmological constant, which warped spacetime in the opposite sense, so that bodies move apart."[6]

The irony is that Einstein's theory — before he "fudged" it — eventually proved correct in one unforeseen way: The universe is not static. It's expanding.

This fact was revealed in the 1920s through observations made by astronomer Edwin Hubble. He noted that in every direction he peered into space, the light emitted by distant galaxies shifted in its wavelength toward the red end of the spectrum. In other words, they all were receding from Earth.

As with relativity, however, the most astonishing aspect of Hubble's discovery was in its implication. With the universe expanding steadily in every direction, all one had to do to was to mentally reverse the process in order to calculate the expansion's starting point. Logic dictated that in the beginning — the beginning of space and time — everything in the universe must have occupied the same point of origin. Then, in an unfathomable burst of creative energy, from that single point the universe came into being.

This primordial explosion that birthed the cosmos has come to be known as the big bang. Considering that this bang gave rise to all matter and energy, and did so from a starting point so small it defies description by modern physics, it makes for a typically human understatement to label it simply "big." However revolutionary, as a scientific theory the big bang has passed the test of independent confirmation. The most dramatic such discovery occurred in the 1960s, when Nobel laureates Arno Penzias and Robert Wilson of Bell Telephone Laboratories detected faint microwave radiation throughout outer space. The discovery confirmed the predictions of fellow physicists, who argued that the massive energy released by a big bang event should still be lurking in space, appearing very much like microwaves.[7]

Science, it seemed, had at last glimpsed the very moment of creation.

Philosophical Objections

Not all scientists considered the triumph of the big bang theory a cause for celebration. Some were troubled that material observations revealed conclusions too similar to religious creation stories. Others were unwilling to accept the notion that discovering the genesis of time and space also established the ultimate limit of scientific inquiry.

Among the most famous detractors of the big bang was the man who helped name it, astronomer Fred Hoyle. According to science journalist Denyse O'Leary, "He hated the theory. As an atheist, he recognized its theological implications, and he didn't like them at all."[8] Hoyle went so far as to team with a few colleagues to promote a counter-theory to the big bang. Dubbed the steady state model, this proposed that as the universe expanded, newly created matter filled in the gaps, so it "would therefore look roughly the same at all times as well as at all points of space."[9] Needless to say, the steady state model did not hold up.

Still, Hoyle's failure does not mean that scientific materialists are now any more ready to consider the big bang an act of God. As we've already seen, there are numerous devices that cosmologists can resort to in order to distance the big bang theory from supernatural causes. The most common method echoes the underlying agnostic principles of Darwinism — dismissing the extraordinary intricacy of the physical universe and its ability to sustain life by portraying these qualities as either the inevitable result of a blind process or else as a lucky accident.

And so modern physicists have proposed a plethora of cosmological models in which supernatural creators are either afterthoughts or ridiculous superfluities. There are cyclical models in which universes are constantly reborn in new configurations, only to collapse and start over again. There are theories of multiple universes which either exist concurrently with our own or pop into existence through odd circumstances, such as the creation of black holes. What most of these theories hold in common is that they are the products of extremely clever imaginations, are usually described with copious amounts of advanced mathematics, and are quite literally unreal.

Indeed, the problem with theories of multiple universes is that they are based more on wishful thinking than fact. The reality is we can observe only one universe, and that universe is exquisitely suited for our existence. What's more, our universe displays a coherence and intricacy on the level that most human beings associate with advanced artistic expression. In the words of physicist and string theory proponent Brian Greene, our cosmos displays an elegance that he can only liken to music.[10]

> **Key Quote**
>
> The reality is we can observe only one universe, and that universe is exquisitely suited for our existence. What's more, our universe displays a coherence and intricacy on the level that most human beings associate with artistic expression.

Now Greene does not advocate intelligent design. He follows in the tradition of Einstein in seeking a unified theory of physics that overcomes the bizarre uncertainties of quantum mechanics and offers a comprehensive explanation for the workings of the cosmos. In this regard, Greene complains that exotic multiple-universe theories, as well as the standard physics model, fail to satisfy the questions humans ask about the only home they know. He writes:

> "Besides its inability to incorporate the gravitational force, the standard model has another shortcoming: There is no explanation for the details of its construction. Why did nature select the particular list of particles and forces …[that make up the cosmos]?
>
> Why do the 19 parameters that describe these ingredients quantitatively have the values they do? You can't help feeling that their number and properties seem so arbitrary. Is there a deeper understanding lurking behind these seemingly random ingredients, or were the

detailed physical properties of the universe 'chosen' by happenstance."[11]

It is well that Greene should wonder at the specific "ingredients" of the universe. The manner in which just the essential forces of nature are fine-tuned leave physicists all agog. Consider, for instance, the strong nuclear force, which bonds the nuclei of atoms in a delicate counterbalance with the weak and electromagnetic forces and their governing of other subatomic particles. The slightest change in the strength of the strong force alone would substantially alter the periodic table. Change it enough and the very elements would fail to form, let alone the stars and planets.

Then there's gravity. Though relatively weak on the quantum level, on a large scale gravity is powerful enough to lock the Earth and its companion planets in stable orbit. At its specific strength, gravity keeps our life-giving atmosphere safely cloaked around us, and ensures the sun maintains the precise size and combustion rate required to provide our planet the correct amount of warmth and energy. Gravity safely pins us to the planet surface without crushing us, yet combines with principles of aerodynamics to let us, in the words of the aviator's poem, slip the Earth's surly bonds.

Fundamental as they are, however, the extraordinary properties of these basic forces are only a minor example of how the cosmos and Earth are arranged in a coherent, life-giving way. Design is woven visibly throughout the fabric of our reality, from galaxies to the construction of our solar system, the Earth and its peculiar moon, a stable yet flexible planet surface, regulated weather patterns and life cycles, genetic coding and even the existence of sentient humans to question why all these properties appear to exist for their benefit.

Given such evidence, it should seem reasonable to conclude that the universe was planned and that humans exist for a purpose. Again, however, materialist science dares to defy such reason by resorting to its favorite catch-all: chance and necessity.

Even if ours is the only universe, the materialist argument goes, it exists according to certain principles. And if these governing principles gave rise to our Earth, then it must be one of a vast number of planets which formed according to the same laws. Consequently, our own life-giving home can only be one of a myriad of similar planetary homes, some with civilizations so advanced they would to us seem god-like.

This argument — often called the principle of mediocrity — attempts to explain away Earth as a mere statistic. But when critics of materialism point out that this theory is based solely on assumptions, the cosmological agnostics often fall back on their ultimate philosophical nullification, a deeply cynical treatment of the anthropic principle. In the agnostic view, the anthropic principle dismisses the way in which the cosmos is apparently designed to suit human life by arguing that it simply had to be that way, otherwise there would be no human life to ask why the cosmos is the way it is. Dawkins, for one, declares that he finds the mediocrity and anthropic principles combine to adequately explain why human life developed on Earth. He writes:

> "There are billions and billions of planets that have developed life at the level of bacteria, but only a fraction of those life forms ever made it across the gap to something like the eucaryotic cell. And of those, a yet smaller fraction managed to cross the later Rubicon to consciousness. If both of these are one-off events, we are not dealing with a ubiquitous and all-pervading *process*, as we are with ordinary, run-of-the-mill biological adaptation. The anthropic principle states that, since we are alive, eucaryotic and conscious, our planet has to be one of the intensely rare planets that has bridged all the gaps [emphasis in original]."[12]

Forgetting for a moment that Dawkins begins his agnostic apology for human life with a complete fabrication — that there are billions of planets with bacterial life — let's consider the philosophical merits of his adherence to the anthropic principle. How intellectually satisfying is it to insist that the cosmos must — without God's help — exist in its present life-favoring form simply because we're here to notice it? For the answer, it's illuminating to recall how atheists such as Dawkins and Dennett so easily and ironically dismiss classical arguments for God's existence such as the one put forward by St. Anselm, who essentially declared that a perfect, holy creator can indeed exist if we are capable of imagining such a being. Frankly, it is intellectually disingenuous, not to say fraudulent, to reject Anselm's argument while employing its underlying logic to make a case for a materialistic cause for the universe. The new atheists simply can't have it both ways. If they are prepared to accept the enigma of their own existence as the mysterious gift of an indifferent cosmos, then they must be willing as well to concede it is reasonable for believers to proclaim the glory of creation as the handiwork of a personal God.

Neither argument, of course — the anthropic principle or Anselm's ontological argument for God's existence — offers complete intellectual satisfaction. They are intended as starting points. The ontological argument, for instance, lays the basis for a developed theology concerning a loving, just, omnipotent Creator, who, although existing outside the time and space he made, nevertheless has a history of sorts through his interaction with humanity. In stark contrast, the anthropic principle can be used to lay the groundwork for a philosophy we've already identified, namely, that the universe exists for no purpose at all, and that human beings, to the extent that they can be said to have free will, are left to derive whatever meaning they care to or can. The consequences of such a philosophy we will examine in detail later. For now it is enough to point out that at least some mainstream scientists have the guts to admit that the anthropic principle leaves them uneasy. Hawking writes:

> "One would feel happier about the anthropic principle, of course, if one could show that a number of

> different initial configurations for the universe are likely to have evolved to produce a universe like the one we observe. This would imply that the initial state of the part of the universe that we inhabit did not have to be chosen with great care."[13]

Greene, as well, complains that the anthropic principle works against his goal of discovering an elegant, comprehensive explanation for the construction of the cosmos:

> "As presented, it is a perspective that is diametrically opposed to the dream of a rigid, fully predictive, unified theory in which things are the way they are because the universe could not be otherwise. Rather than being the epitome of poetic grace in which everything fits together with inflexible elegance, the multiverse and the anthropic principle paint a wildly excessive collection of universes with an insatiable appetite for variety. It will be extremely hard, if not impossible, for us ever to know if the multiverse is true."[14]

'Privileged Planet'

Finally, there is an aspect to the debate over both the mediocrity and anthropic principles that mostly has been overlooked by scientists — until now. Raising their objections in a recent and controversial book, *The Privileged Planet: How Our Place in the Cosmos is Designed for Discovery*, authors Guillermo Gonzalez and Jay W. Richards point out that it is not only remarkable human beings observe that they occupy a hospitable planet in a finely tuned cosmos — it is equally remarkable that human beings can make these observations at all.

The argument forwarded by Gonzalez, an astronomer, and Richards, a philosopher, defies mainstream thinking in each author's respective field, but holds in its favor two important aspects: It appeals to what most human beings feel intuitively about their existence, and it conforms to the known facts.

In summary, Gonzalez and Richards suggest that many of the qualities that contribute to Earth's habitability also contribute to the development of technology and science. What's more, many of these qualities are out of keeping with common patterns in the cosmos. That these qualities exist individually in a universe such as ours is improbable; that they exist together to the benefit of a single planet such as Earth is highly improbable.

As the authors put it, Earth holds such a special place in the cosmos that it defies explanation by mere material causes:

> "Even more mysterious than the fact that our own location is so congenial to diverse measurement and discovery is that these same conditions appear to correlate with habitability. This is strange because there's no obvious reason to assume that the same rare properties that allow for our existence would also provide the best overall setting to make discoveries about the world around us. We don't think it's merely coincidental."[15]

Key Quote

Gonzalez and Richards suggest that many of the qualities that contribute to Earth's habitability also contribute to the development of technology and science. What's more, many of these qualities are out of keeping with common patterns in the cosmos.

To support their thesis, Gonzalez and Richards provide a comprehensive analysis of Earth's special qualities best encountered in their book. A few of their points lend themselves to summary here.

The Earth's remarkable good fortune, for instance, begins with its position in the Milky Way galaxy, its host star, and the planetary company it keeps.

As for its galactic position, the Earth is extremely well-favored. By being located between outer spiral arms, our home planet avoids the destabilizing gravitational forces of an apparent black hole in the galaxy's center, as well as large astral bodies there. The galactic center is also subjected to much greater amounts of radiation. What's more, the Earth's dust-free location toward the outer rim of the Milky Way not only allows for an excellent view of most of our own galaxy, but for the viewing of deep space objects as well.[16]

Few stars, meanwhile, could be more suitable for sustaining life than our sun. As a source of energy, it is both powerful and highly stable. Gonzalez and Richards point out that the "Sun is among the 9 percent most massive stars in the Milky Way galaxy." And yet its light output varies by only a fraction of a percent.[17]

The planetary make-up of our solar system is remarkable as well. All the planets around the sun maintain stable orbits and do not interfere with Earth's own trajectory. Among the outer planets are the gas giants Saturn and Jupiter, which function as sentinels of sorts, their massive gravity diverting comets, asteroids, and other space debris which otherwise might enter the inner solar system and collide with Earth.

Besides benefiting the Earth in a physical sense, however, our system's fellow planets aid our ability to understand basic principles of science. It was this ability to progressively study the movements of nearby heavenly bodies that helped scientists first propose and then confirm various foundational concepts of mathematics and physics. Greatly benefiting this study is the fact that the planets of the solar system are within view of Earth and complete their orbits of the sun in substantially less time than the normal human life span.

Again, as Gonzalez and Richards point out, "Even geniuses like Kepler and Newton needed a planetary playpen to discover the laws of motion and gravity and to realize that they apply throughout the cosmos."[18]

More astonishing than our view of the planets, however, are the latest facts that support our sense that the solar system is improbably favorable to life and technology. These stem from the recent discovery of planets outside our own solar system — a discovery that so far reveals no clear pattern for the construction of planetary systems. Indeed, so far scientists searching for extra-solar planets have only identified situations very unlike our own hospitable array — mostly massive bodies orbiting too close to their host stars to be habitable. In the future, scientists may refine their search techniques to allow for the detection of Earth-like planets, but for the present the conclusion remains that Earth is undeniably unique.

Indeed, the Earth is not only unique in its sun and fellow planets, it is favored by its remarkable habitat. As for its atmosphere, the Earth is cloaked by a complex collection of gases which simultaneously collect life-giving solar energy while blocking harmful radiation. The atmosphere also helps regulate Earth's climate to relatively moderate levels. And it is transparent enough to offer the single most crucial factor to the development of astronomy — a clear view of the night sky.

The Earth is also benefited by its remarkable moon — one that is disproportionately large compared to the solar system's many other moons. Our moon helps keep the Earth in stable rotation, and maintains ocean tides, serving to further regulate weather patterns and the crucial carbon cycle. Yet the moon also functions as a boon to science through its most improbable quality — the fact that, due to its size and position, it appears in the Earth's sky at almost the exact size of the sun. This leads to the phenomenon of perfect solar eclipses, which further allows for important scientific studies, including the famous observation of the 1919 solar eclipse which helped confirm Einstein's theory of relativity.[19]

All in all, the extraordinary qualities of Earth lead Gonzalez and Richards to an equally extraordinary conclusion, that, "habitable environments are exceedingly rare. The fact that they are also the best overall places for scientific discovery forms a relatively independent pattern. So we have good reason to suspect that things have been intentionally arranged, even if this came about through the interaction of natural laws and initial conditions."[20]

In other words, the fact we exist and can recognize the cosmos for what it is can only mean one thing — the universe and its inhabitants are the product of design.

Chapter Eight

Of Miracles and Math

Why Two Plus Two Equals The Mind of God

"The illusion of design is a trap that has caught us before, and Darwin should have immunized us by raising our consciousness."

— Richard Dawkins, *The God Delusion*[1]

"The process of mathematization has uncovered intimate connections between mathematics and science that reveal the marvelous coherence of creation – something for which we can glorify God."

— Editors, *Math in a Postmodern Age*[2]

The new atheists are not blind to the fact that the material world exhibits design. Their contention is that this design should not be attributed to an existential supernatural agent.

Here, then, lies the crux of the matter. Much of what both theists and atheists observe and can describe concerning physical reality they agree upon. The difference — and oh, what a difference — is in the interpretation. Atheists dismiss supernatural explanations because — with apologies to Will Rogers — they never met a deity they didn't disparage. Disgusted by religion and its creeds, they prefer to construct an ethos based on the secular scripture of science.

And yet even science requires that certain fundamental constructs be accepted by faith. Information, finely tuned fundamental forces, the fact that the cosmos behaves in a predictable manner — all demand an ultimate source of rationality and order greater than the cosmos itself. Raw matter and chance cannot possibly account for such intricacy; by definition an extra-cosmic source of order and information must be supernatural.

Consequently, science does not necessarily have the agnostic cast most scientists would have us believe.

Words and More Words

As with many purveyors of faith, however, when the new atheists find their arguments failing from lack of evidence and logic, they rebut their critics with a device at which they remain particularly adept — empty rhetoric. By employing far-fetched yarns to buttress a materialistic apology for phenomenon that reasonable people would admit lie beyond the scope of human understanding, the new atheists ironically undermine the very thing at the core of their worldview — science itself. Indeed, far from liberating their fellows from what they call the pernicious influence of religious superstition, the new unbelievers actually attack rationality by touting belief in a mechanistic universe that does not necessarily abide by its own mechanisms. According to their lights, we live in a wondrous cosmos fraught with miracles that no one must admit ever happened.

> **Key Concepts**
>
> New atheists try to boost their philosophy by claiming its foundations are as credible as those supporting mathematics. The problem is:
> ❖ Even math is plagued by irrationalities and logical fallacies.
> ❖ Math can be employed as a metaphor for religious faith because it is something abstract that nevertheless applies to physical reality.
> ❖ Some academicians see math as a means by which God works in creation.

Dawkins, for instance, agrees that phenomenon such as the emergence of life and human consciousness defy explanation and are too improbable to have just randomly occurred.[3] This does not mean he is ready to ascribe their existence to some supernatural force. No, he insists that Darwinism — particularly the concept of natural selection —– can easily account for such things.

Indeed, Dawkins declares that by breaking down improbable phenomena into a succession of slightly less improbable things, natural selection is capable of achieving nearly anything. Life origins, complex living systems, genetic coding, even human morality can all be explained by Darwin's central idea — one that Dawkins venerates with an almost mystical awe.

As Dawkins insists, "A deep understanding of Darwinism teaches us to be wary of the easy assumption that design is the only alternative to chance, and teaches us to seek out graded ramps of slowly increasing complexity."[4]

And again, when accounting for all living things: "Natural selection is a real solution. It is the only workable solution that has ever been suggested. And it is not only a workable solution, it is a solution of stunning elegance and power."[5]

Caught up in his Darwinian raptures, however, there are a number of things Dawkins fails to make clear. He does not explain how natural selection can be described as elegant when it is also said to be blind and purposeless. He does not mention scientists such as Gould, who, as we have already noted, do not necessarily believe that evolution automatically moves toward greater complexity. And he omits the fact that his ill-defined concept of natural selection is, like Einstein's infamous "fudge factor," merely a convenient device to surmount the unpleasant statistical obstacles to Darwinian evolution. Indeed, Dawkins' claim that natural selection makes any particular evolutionary feat just probable enough ignores the reality that for living things to evolve requires not just one unlikely event, but a series of unlikely events — where the odds against successful completion increase exponentially with every successive obstacle.

Dennett, as well, employs vague Darwinian principles in attempting to account for religion as natural phenomenon. Rejecting out of hand the idea that one faith, at least, might be founded on wisdom that transcends material reality, ("it simply isn't true")[6] Dennett instead speculates on how evolving memes could have transmitted religious ideas through "various continuities" including the "parent-child instructional pathway."[7]

Dennett illustrates the aforementioned pathway with a Darwinian fable about how human adults learned to love their babies. He writes:

> "...human beings are irresistibly captivated by the special proportions of a 'baby face.' ...It is not that baby faces are somehow intrinsically darling (what on earth could that mean?) but that evolution hit upon facial proportions as the signal to trigger parental responses..."
>
> "We don't love babies and puppies *because they're cute.* It's the other way around: we *see them as cute* because evolution has designed us to love things that look like that." (Emphasis in the original)[8]

Remember, of course, that when Dennett declares evolution "designed" us, he doesn't mean in the same way an actual designer would have, because that sort of blundering admission would overturn his entire premise. Indeed, it's hard to determine exactly what Dennett means by presenting such a just-so story in the guise of science. What evidence has he to confirm that baby faces trigger some intrinsic adoration mode in human adults? Is there a gene that regulates this baby-loving compulsion? Perhaps a neurotransmitter for inducing adults to produce nonsensical burbling sounds that infants seem to enjoy hearing and imitating?

The evidence from social history certainly undercuts the idea that adults can't help but dote on their infants. Infanticide has been a fairly common practice throughout history and is still advocated in some quarters.[9] Child abandonment, as well, remains problematic even in the most developed societies, with fathers, in particular sometimes displaying notorious levels of neglect in regards to their own offspring.[10]

Then there's the paradox of "fitness" presented by the idea that love for children slowly evolved as a triggered response. One can't help but ask: How did early generations of human infants — cute or otherwise — manage to survive when their parents had yet to be "designed" by evolution to adore them? And what of the babies — dare we say it, poor dears — who are just plain ugly?

Frankly, it's absurd to suggest that the relationship between parents and their children can be characterized in such a simplistic way. The fact is even cute babies are often trying to care for. Yet adults lovingly raise them for a number of complex reasons: out of a sense of duty and hope, and out of a desire for progeny — which often includes the wish to pass on a particular ethos, including a religious faith.

The point here is not so much to rebut one particular conjecture of Dennett's regarding social evolution, but to illustrate how rejecting the idea of a divine creator necessarily leads to materialistic falsehoods and pseudoscience, strata upon sinking substrata. Ironically, the new atheists could see this if they didn't shut their minds to the possibility that God is real and instituted the very systems they study. Blinded as they are, they fail to detect the supernatural design evident in the most fundamental discipline underlying the studies they so claim to revere.

Playing the Numbers

Probably no human endeavor encapsulates the conceit that the universe is orderly and can be described in a rational manner better than the study of mathematics. In the Western tradition, where mathematical concepts are first confirmed by intellectual "proofs" before then being tested via practical application in the physical world, simple math equations are often upheld as ultimate, incontrovertible fact. Two plus two, we are told, have always equaled four, and always shall, world without end. Amen.

Dennett invokes math to illustrate why he rejects religion in favor of science, particularly in the advanced calculations used to confirm theories in physics:

> "There is a big difference between religious faith and scientific faith: what has driven changes in concepts in physics is not just heightened skepticism from an

> increasingly worldly and sophisticated clientele, but a tidal wave of exquisitely detailed positive results — the sorts of borne-out predictions that [Richard] Feynman printed in defending his field. … You can build something that depends for its safe operation on the truth [of physics] and risk your life flying it to the moon."[11]

One could ask how it is that Dennett maintains such confidence given the fact that history is littered with discarded scientific theories. Or how he can laud the exquisite positive results of physics when many astronauts launched into space have failed to return in one piece. But again, the aim here is not to denigrate physicists or mathematicians, but to point out the duplicity in the thinking of materialists such as Dennett.

Dawkins and Dennett, for instance, insist that nothing in the universe can be true unless science confirms it. God, whether creator, judge or redeemer, lies beyond the scope of physical observation, so God cannot exist. Moral truth, sin and repentance, have no real meaning beyond what biologists and sociologists can observe and tabulate. If something cannot be described with numbers, ratios and formulae, it must remain forever nebulous and dubious.

Lack of Certainty

Which brings us to a major problem, one so fundamental there is hardly any excuse for Dennett not to have mentioned it. He praises the reliability of science, and by association its handmaiden, math, but as a philosopher he must know that math's intellectual underpinnings nowadays are nothing less than a shambles. The idea that math stands as a paragon of logic collapsed decades ago after some of the greatest minds tried — and failed — to construct through pure reason a coherent structure of proofs. Simply put, mathematicians no longer believe their ideas are demonstrably true. They still perform great feats of calculation and help design the kind of exquisite contraptions so admired by Dennett — machines that continue to amaze (with a few spectacular and fatal exceptions.) But as for why their calculations work, they cannot say.

This disconnect between math's practical application and its intellectual foundations springs from more than petty and esoteric distinctions. It has to do with the most basic understanding of reality and how we perceive it. To better illustrate it is helpful to consider examples from math history and philosophy.

Interestingly, the tension that arises from the paradoxical nature of math can be traced to its first great practitioners, the ancient Greeks. Followers of Pythagorus, for instance, were among the first to view the world as rational and orderly to the extent that it could be examined in what we now call a scientific way. Euclid, with his treatise on geometry, gave the world its paradigm for an axiomatic approach to studying and documenting abstract knowledge. The internal logic of Euclid's proofs and his precisely defined terms, as we've noted, are still upheld as the archetype of good mathematics and evidence that truth can be gleaned from observing the cosmos.

But unlike the Pythagorians, who regarded math with a sort of mystical awe to the point that they revered numbers as elemental properties of nature itself, later Greeks came to view mathematics as merely a tool for getting at a deeper reality. In Plato's view, as part of our imperfect material existence, math at best could only represent shadows of the ideal world where alone resided perfect truth and beauty. Math was a useful abstraction, but nothing more. As math historian Morris Kline writes, "From the standpoint of the search for truths, it is noteworthy that Ptolemy, like Eudoxus, fully realized that his theory was just a convenient mathematical description which fit the observations and was not necessarily the true design of nature."[12]

Not until after the Renaissance did mathematicians finally shed this Platonic reticence and embrace a bold new outlook that opened the way for full-fledged materialism. Ironically, this philosophical shift was due in great part to the theistic reverence of some of the great Christian scientists. Awestruck by the discoveries of "laws" by which they could predict the movements of the heavenly bodies, men such as Copernicus, Descartes and Newton were inspired, as we've noted, to view the cosmos as a harmonious mathematical machine, wound up and hung in place by God. As Kline points out, however, this newfound ardor for the explanatory power of math led these scientists to "emphasize quantitative mathematical laws as opposed to physical explanations."[13] This in turn gave rise to the fallacy that being able to describe a phenomenon in terms of mechanics was an adequate substitute for actually understanding the physical characteristics of the phenomenon. The rise of secularism in the so-called age of reason compounded this mistaken emphasis. Further advances by men such as LaPlace not only diminished a place for God in the physical realm, it prompted leading intellects to champion mathematics as the ultimate means for comprehending reality — a reality increasingly described in mechanistic terms.

Then came the unthinkable. Mathematicians uncovered bizarre manifestations of their craft that, while true, defied conventions that had been held as unassailable. Worse, by the turn of the twentieth century attempts to reconcile these enigmas only served to reveal that math, however useful, could not be reduced to a neat set of axioms. Ultimately, math simply was not logical.

Examining these conundrums requires delving into the esoteric and, in some cases, the absurd. This descent into the intellectual netherworld presents a worthwhile venture, however, since it offers a glimpse into the puzzles that still addle the minds of some of our greatest thinkers.

We begin with nothing less than the overturning of Euclid. By the early 1800s, the German mathematician Carl Friedrich Gauss postulated something which the founder of geometry had declared impossible — triangles whose angles added up not to exactly 180 degrees, but sometimes more, sometimes less. Gauss had discovered what is now its own branch of geometry, based not on the classic (and mostly imaginary) flat plane, but on curved space.[14]

More shocking than the discovery that new kinds of geometry existed, however, was the revelation that Euclid's axiomatic method was faulty. And the problems extended to other branches of math. In the field of algebra, for instance, a conundrum arose involving complex numbers known as quaternion. Evidently, when two quaternion are multiplied, the result differs depending on which number is written first in the equation — an outcome which calls into question basic assumptions about number theory itself.[15]

And as we have noted, attempts to restore math's foundations by reconstructing a more rigid set of axioms didn't help. Men such as Bertrand Russell learned that reason could only take them so far, that even the best system of logic ultimately faltered when confronted by an insurmountable paradox. In Russell's case, his work in set theory reached its limit when he posed the now-famous question: "What is the set of all objects that belong to no set?" Proof that riddles such as his are not mere rhetoric came with the rather dour pronouncement of mathematician Kurt Gödel in the 1930s. As math historian Leonard Mlodinow writes, Gödel "proved that in a system of sufficient complexity, such as the theory of numbers, there must exist a statement that cannot be proved either true or false. A corollary of Gödel's theorem is that there must exist a true statement that cannot be proved."[16]

As one can imagine, assertions such as these did little to ameliorate the brewing controversy over the foundations of math. In fact, by the time Gödel insisted that a lack of certainty was intrinsic to mathematics, mathematicians themselves had divided into various philosophical camps that were, according to Kline, "at war with each other."[17]

Among these factions arose a school of thought which actually is quite un-mathlike, but nevertheless represents an important change in the foundations for human reasoning in that it returns to a kind of Platonism. This particular view argues that there are aspects of math that can never be proven through logic, but which we can accept as true through experience or sheer intuition. This argument, of course, frustrates those who wish that math at least, among all the things that human beings struggle to know, could be precise, coherent, indisputable. Yet even the most ardent student of logic finds himself thwarted by some of the most basic aspects of math. Infinity, for example, is invoked in many of its branches, from number theory to geometry. But is infinity real? Kline quotes Jules Henri Poincaré as saying, "What we call infinite is only the endless possibility of creating new objects no matter how many objects always exist."[18] If this is so, do we redefine the concept of the infinite as merely referring to large, indeterminate but ultimately finite quantities, or do we confess to the possibility of timeless, boundless things existing beyond our own realm?

Key Quote

The great paradox of math is that it hints at wonders beyond what is possible in our finite universe, beyond even what our finite minds should be able to conceive.

Here is our difficulty. As a mere model or description of reality, math can never be equal to the thing it describes because then it would have to be the thing itself. Likewise, since math is an intellectual endeavor by humans who also are only part of the physical reality they inhabit, math as humans know it can never be greater than the thing it describes because then it would be supernatural. And so the great paradox of math is that it hints at wonders beyond what is possible in our finite universe, beyond even what our finite minds should be able to conceive.

Again, the notion of math as intuitive truth may seem onerous to those who wish to inhabit a mechanistic cosmos whose working hold no clues to a higher reality. But as Kline warns, those who wish to live in a universe that functions strictly by the numbers risk living in a hollow realm of make-believe. He writes:

> "Modern science has been praised for eliminating humors, devils, angels, demons, mystic forces, and animism by providing rational explanations of natural phenomena. We must now add that modern science is gradually removing the intuitive and physical content, both

of which appeal to the senses; it is eliminating matter; it is utilizing purely synthetic and ideal concepts such as fields and electrons about which all we know are mathematical laws. Science retains only a small but nevertheless vital contact with sense perceptions after long chains of mathematical deductions. Science is rationalized fiction, rationalized by mathematics."[19]

Profoundly Mysterious

Without God, then, there is in some sense no real reason why math should exist and function the way that it does. Unlike life forms and heavenly bodies — which we've considered in some depth — math is utterly abstract, incorporeal, in a sense even more of a paradox than the physical things in the cosmos, because although math is fundamental to our ideas about physical order and structure, as far as we can tell math only exists in the minds of humans.

Think about it. There really are no such thing as numbers. No one has ever seen, heard or felt the number four, for instance. And though one of the first things we teach our children is to count objects in sets, the truth is there is no such thing as just four or ten or one million of anything. There are merely subsets of larger sets of the whole set of every particle in existence that we organize in our minds for the sake of clarity. (And if we set out to count all the particles in the universe, physicists tell us the number would be constantly changing as some matter gets sucked into black holes.) What's more, the further a student advances in the study of math the more he encounters principles that in terms of the physical

> **Key Quote**
>
> Despite what Dennett and his compatriots claim, faith in math is very much akin to faith in the supernatural. Even the ancients realized that unless those who study math recognize it as the manifestation of a greater, transcendent source of structure and order, it remains a hollow, perverted faith.

realm amount to nonsense, such as lines stretching into infinity, negative and imaginary numbers.

And so, despite what Dennett and his compatriots claim, faith in math is very much akin to faith in the supernatural. As we've already noted, even the ancients realized that unless those who study math recognize it as the manifestation of a greater, transcendent source of structure and order, it remains a hollow, perverted faith.

As one math historian puts it:

> "…a high regard for mathematics was nurtured by Greek sources and moved out to become an established feature of Western intellectual culture. This program of mathematization shared many characteristics with religious belief — it involved faith and commitment and assured the existence of something whose nature did not depend on anything else, but which determined the nature of other things."[20]

Math, then, by virtue of its reliability has indeed proven a very real thing. But is it real enough, even as the foundation for science, to serve as a means for resolving the most important quandaries that confront human beings? We can be extremely confident, for instance, that six times two equals twelve, but where does this get us when trying to determine if it is ever ethical to take the life of another human? Should we, like the utilitarian disciples of Jeremy Bentham, attempt to reform social policy based on a scientifically compiled human efficiency graph cross-indexed with a happiness matrix? Common sense and history tells us that such an approach is preposterous.

Aside from its disconnect from true human values, there is another reason why we should distrust math when it is used as a basis for determining how people should live. Because like the various branches of science it supports, math can be distorted to support dubious philosophies.

Consider again how materialists such as the new atheists often abuse math to prop up their worldview. We've already discussed how Dawkins ignores mathematic principles by arguing that natural selection somehow negates the rule that if a particular desired outcome requires more than a handful of successive random events, the odds of it occurring are not even worth discussing. Many fellow Darwinists similarly distort probability theory by implying that time and chance are capable of delivering just about any biological outcome.

But in itself chance has no creative power. Similarly, probability can provide no real explanation for why or how things occur; it is merely a mathematical tool for describing very complex problems. It offers context without negating the fact that everything has a cause and effect. (Thus, sentient life doesn't just randomly occur any more than two plus two just happen to make four.) To use Dennett's own analogy, scientists don't plan a manned rocket trip based on the notion that certain physics equations happen to be correct once in a while, or even more often than not.

> **Key Quote**
>
> In math, we have an abstract manifestation of something hard and real and true that exists outside of our universe, a sort of outward visible sign of an omnipotent spiritual grace.

Given this latter fact, we find ourselves having come full circle in the question of why the cosmos displays order to the degree that it can be understood with mathematical precision. The new atheists certainly offer no substantive explanation. Dennett, for instance, recognizes the absurdity of classifying mathematic principles as mere memes subject to constant mutation. Of course, for him and his compatriots this raises an embarrassing paradox. If mathematics aren't memes, they are not subject to evolution, which means they lie outside the laws that some ultra-Darwinists claim govern all of material reality. According to this view, math must be supernatural.

Language of God

But numbers aren't gods; neither is arithmetic holy writ. Again, it would seem that the ancient Platonists were indeed on to something: In math, we have an abstract manifestation of something hard and real and true that exists outside of our universe, a sort of outward visible sign of an omnipotent spiritual grace. In the intangible realm of numbers and equations we may indeed be glimpsing the language of God.

If this idea strikes some post-modernists as too mystical, it is helpful to consult history. In the Western tradition, especially, many great scientific minds apparently accepted the premise that mathematical truth proceeded from God. Though many of these, such as Newton, Kepler and Galileo Galilei are considered responsible for contributing to the ascendancy of science over faith, they themselves did not adhere to the modern notion that the two realms of knowledge need be segregated.

To again cite the historians from earlier in the chapter:

> "Like Kepler, Galileo's mathematical viewpoint of the world was grounded in his religious and philosophical orientation. Galileo took a traditional Augustinian viewpoint on the source of natural and scriptural revelation — God is the acknowledged Author of both the book of Scripture and of Nature. Therefore, divine truths revealed by either one cannot contradict those of the other, though human interpretations of each might give rise to conflicts."[21]

And despite the atheistic outlook that dominates much of modern science, there are still some thinkers who are not afraid to ascribe the existence of math to a divine source. One current philosophy, known as theistic activism, argues that an existential God essentially thinks into existence the intangible truths of numbers, formulae and other aspects of math:

> "Roughly, they are God's thoughts, concepts, and perhaps certain other products of God's 'mental life': This divine activity is thus causally efficacious: the abstract objects that exist at any given moment, as products of God's mental life, exist *because* God is thinking of them; which is to say that God creates them." (Emphasis in the original)[22]

This philosophy goes on to assert that God conceives of these things merely because it is in his nature to do so. "It is God's power that sustains abstract objects. It is, moreover, not possible that God withhold that power."[23]

Chapter Nine

NOT GOOD ENOUGH
The New Atheists Are Appalled By Evil But Can't Explain Why

"The problem of vindicating an omnipotent God in the face of evil ... is insurmountable."
— Sam Harris, *The End of Faith*[1]

"In a sense, [belief in God] creates, rather than solves the problem of pain, for pain would be no problem unless, side by side with our daily experience of this painful world, we had received what we think a good assurance that ultimate reality is righteous and loving."
— C.S. Lewis, *The Problem of Pain*[2]

This much we've established: The universe displays intricate design, and functions with mathematical precision. What's more, math itself, with its internal logic and elaborate proofs, exhibits a coherence so reliable that some thinkers have attributed its abstract characteristics to the very mind of God.

Which leads us to the most vexing of all metaphysical and philosophical conundrums. The order and precision we encounter when observing the cosmos may indeed open our minds to the possibility of a Creator, but then the problem becomes determining just what sort of God we're dealing with. The elegance of physics and other mathematical applications might inspire mild reverence for some grand architect, but in a very practical sense — in terms of life and death, the human struggle for existence so famously characterized as one of quiet desperation — it would help a great deal to know whether the God who made the universe, with all its wonders and terrors, is to be scorned, appeased, or simply ignored. In short, how does the grandeur of the cosmos square with the disease, decay, violence and death that daily confront every living creature on our lonely world? For the new atheists, the answer is simple. God cannot be reconciled to the problem of evil. The solution is to scorn him out of existence.

Their argument goes something like this: There can be no benevolent Creator because the universe is nowhere near benevolent enough. Nature is cruel, and so are its creatures, especially the only species whose members claim to be rational enough to determine good from evil. This is especially true when it comes to religious people, whose hypocrisy exposes the falseness of their creeds. Religious people not only fail to do the good they say they should do, but often inflict evil instead.

Lastly, the new unbelievers spurn the notion that human free will accounts for the possibility that evil can exist without tainting the goodness of the Creator who made that free will.

What we shall see, however, is that these accusations are utterly spurious. They represent the culmination of all that is false in the new atheists' interpretation of science, philosophy, and human nature. Worse, by the invective they hurl at one religious tradition in particular, they betray how much of their disbelief is fueled by frank bigotry.

> **Key Concepts**
>
> New atheists insist no good God could exist as long as evil exists. This argument fails to hold up given the fact that:
>
> ❖ The new atheists don't blame just any god for failing to restrain evil. They blame the God of the Bible.
>
> ❖ Rejecting a divine source of morality leaves them without absolute grounds for determining right from wrong.
>
> ❖ Attempting to form a moral code based on human consensus yields dubious results.

When God Disappoints

As we've noted, one novel thing about the new atheists is the curious way in which they employ the problem of evil against theism. On one hand, they avoid the typical materialist dodge of claiming they have no objective standard for calling anything evil. On the contrary, they tally the alleged sins of religion with gusto and indignantly denounce the calamities of nature as if they were unforgivable crimes.

Dawkins goes so far as to call the God of the Bible a "monster." He then mocks the biblical deity with the tired logical fallacy that purports to show God's claims to omniscience and omnipotence are a sham: If God knows ahead of time when he'll intervene in nature, he's powerless to stop himself from intervening. And yet, implies Dawkins, it's apparent that God fails to intervene for the good of his creatures because so many suffer.[3]

Again, it is curious that in seeking to dismiss the idea of a benevolent deity the new atheists seem primarily to consider the God put forward by Judeo-Christian tradition. The aforementioned authors spend little time railing against dualism or paganism, for instance. However marginalized by monotheism, these two belief systems at least offer philosophical alternatives to the classical problem of evil. In dualism, a god might transcend the constant warring of good and evil without countenancing either. And paganism, with its crowded pantheon, allows for the source of vice and virtue to be divvied up between any number of lesser gods. Nevertheless, Dawkins and his colleagues seem obsessed with the God of the Bible. Journalist David Aikman also notes this preoccupation of the writers we've discussed, whom he calls, in a glib reference to St. John's Apocalypse, "the Four Horsemen." He writes:

> "The overwhelming impression one gets from reading the Four Horsemen is that they are asserting the non-existence of someone they sort of know — or at least think they know *about* — but whom they dislike venomously (Dawkins), clandestinely admire (Dennett), or simply would not care to become acquainted with if he did exist (Hitchens and Harris.)" [emphasis in the original][4]

Perhaps the new atheists spend so much energy denouncing the supposedly nonexistent God of the Bible because they feel compelled to defend their own standard of benevolence, however abstract it may be, against the goodness that Yahweh extends to his people and demands in return. That the new atheists should make this comparison is odd for several reasons. The first is that Christian doctrine has much to say about the existence of evil and how God has gone to great lengths to redeem his creatures from it. But the new atheists reject both parts of this doctrine out of hand. They deny the biblical definition of evil — rebellion against God — then ridicule the idea that man requires redemption from this wretched state. This puts the new atheists in the awkward position of explaining where — as materialists — their standard of goodness comes from in the first place.

This problem of accounting for goodness in a godless universe goes beyond the establishment of cultural ethics. When they're being honest, religious and nonreligious folk will admit that people rarely reach complete agreement on how best to live together — and that they often fail to live up to rules that are agreed upon. No, this problem runs as deep as trying to comprehend human nature itself, because it is intrinsic in all human beings to feel that some things are just and other things are simply wrong. And this question is not one that science, Darwinism especially, is equipped to explain.

> **Key Quote**
>
> Perhaps the new atheists spend so much energy denouncing the supposedly nonexistent God of the Bible because they feel compelled to defend their own standard of benevolence, however abstract it may be, against the goodness that Yahweh extends to his people and demands in return.

Even on a philosophical level, as we've already discussed, evolutionary theory cannot escape the damning paradox that any creation system that brings about a universe where evil exists must be culpable for that evil. Even Harris grudgingly admits, "The perverse wonder of evolution is this: the very mechanisms that create the incredible beauty of diversity of the loving world guarantee monstrosity and death."[5]

Natural Immorality

More perverse have been attempts to establish moral systems based purely on evolutionary theory. Indeed, if nature is only concerned with selecting those organisms most fit for propagating their kind within a certain environment, then any behavior exhibited by such organisms must be interpreted in light of how that behavior assists them in that goal — surviving and reproducing. And this is just what we see in recent history.

Darwinism, in fact, has been exploited to justify all sorts of appalling values and behavior. It has been used to excuse racism and other types of discrimination, cruel experimentation on human beings, sexual profligacy and many other evils. For instance, author John West points out a recent book by American authors Randy Thornhill and Craig Palmer that explains rape as the result of how sexuality evolved in human males and females.[6] In other words, if Darwinism is all we have to explain behavior, then any behavior that advances the aims of Darwinian evolution, however vague, cannot be condemned.

This historical and philosophical difficulty with Darwinism should deter the new atheists from excessive moralizing, but it doesn't. They rant about evil, whatever they deem it to be, and how human beings should disdain it, but they can't fully explain what goodness is or how to be good.

Dennett, for instance, argues that evolution produces good behavior such as cooperation as part of an overall survival strategy. He suggests that organisms — beginning with the most primitive, of course — learned through trial and error that working together allowed for more positive outcomes than constant competition. He then cites computer models based on mathematical game theory which purportedly show how cultural mores could evolve from such humble beginnings — results achieved, we might add, thanks to the work of designers who built in parameters meant to control random factors. Of these models, he writes:

> "They show conditions under which, bucking the constant headwind of evolution's myopia, organisms can come to be *designed* by evolution to cooperate, or more precisely, designed to behave in such a way as to prefer the long-term welfare of the group to their immediate individual welfare."[Emphasis in the original][7]

Now, for the caveats. Bear in mind that when Dennett speaks (again) of organisms being "designed," he doesn't mean this in the normal sense of them being crafted by an intelligent creator (as we've already noted, that would make him a creationist). Nor does he necessarily believe that creatures can display true altruism (that would make him religious). To begin with, Dennett argues that philosophically it is difficult to declare any behavior truly selfless because the person engaging in the behavior may be inspired to selflessness for reasons that are self-gratifying. What's more, because Dennett analyzes everything through the filter of evolutionary theory, he insists that altruistic acts must be interpreted as behavior intended to gain some other benefit in terms of survival or reproduction.[8]

This, again, is how Dennett concludes that human morality and religious creeds can be described as evolved, material things that do not represent transcendent truth. Thus, God, the Ten Commandments, the Golden Rule, heaven and hell can all be disregarded as passing fancies. Or more specifically, they are memes, independently evolved "cultural recipes" that latch onto human brains (and not necessarily the other way around) in their own struggle to propagate and survive. Memes, Dennett informs us, can be harmful, neutral or beneficial.[9] He does not tell us how to identify which category individual memes fall into, except to assure us that probably all religious memes are bad.

Again, some crucial qualifiers. First, it should be pointed out that meme theory is by no means orthodox science. As Aikman puts it, "...the 'meme' is an entirely speculative entity. No scientist has ever found a way to observe it or measure it, much less reproduce its likeness in a laboratory setting. It is an alluring theory of cultural change, but its existence has never been proved."[10]

Then there's the fact that Dennett is unclear on whether certain other "cultural recipes" (aside from the ones he venerates) can be classified as objective truth simply awaiting discovery. We've already considered Dennett's high regard for math. In another instance he speaks of "arithmetic, and many other timeless and absolute systems of truth."[11] How are we to interpret such a declaration, given Dennett's ardent attachment to Darwinism? Is it possible that some religious truth is as timeless and absolute as math and physics? Are humans simply revealing inalterable realities when they declare that violence begets violence, that all human life is sacred, that nature is so corrupted by sin that divine grace is the only thing that keeps it from spiraling into chaos?

If we consider these things as mere memes, there is no way of knowing. But of one thing we can be sure: Dawkins, Dennett and his cronies will be more than willing to offer their services as secular priests to guide us in our muddled musings.

So-So Sociology

In addition to evolutionary theory, Dawkins and Dennett are fond of referring to research from the social sciences in attempting to define a human code of morality. Their aim is to prove that morals, especially those shared more or less universally, could have developed without being revealed by divine edict.[12] Their success in this aim is less than complete, as one might expect from any endeavor whose principal source of data comes from people toying with the minds of others.

Consider an experiment cited by Dawkins, one studying the choices of individuals asked to ponder a hypothetical life-and-death crisis regarding a runaway railroad car. Study participants were asked if, given the choice, they would allow the hurtling car to continue on a course which would kill several people, or if they would flip a switch and divert the car onto another track which would result in the death of only one. The study then proceeded to ask participants if they would push a bystander in front of the hurtling railroad car in order to halt it and spare the lives of many others.

From there, the choices offered to people were made more perplexing and morally dubious. Starting from the relatively antiseptic decision involving throwing the railway switch, participants were finally asked if it would be acceptable to kill a healthy visitor at a hospital in order to harvest his organs and save the lives of five other people.

Dawkins insists that the fact most study participants offered the same answers — it's better to save many lives than one, but actively taking another human life is repugnant — proves natural forces can establish moral universals. He adds that this conclusion is especially true because the majority view in the study was shared by both religious and nonreligious participants.[13]

But aside from reassuring railway patrons that they aren't in any real danger of being pushed in front of runaway trains, what do these kinds of studies really tell us? When it comes to establishing something as fundamental as the rules for human interaction, how much can we rely upon the way people respond to contrived situations designed to elicit responses within an artificially narrow range? According to critics of the social sciences, the answer is — not much.

The problems with studies of this kind are numerous. To begin with, they are usually entirely hypothetical (as in the hurtling railway car) or set up as a sort of laboratory game. In either case, there is little, if anything, at stake, so it is unlikely that the person asked to make a decision can feel the full weight of the possible consequences.

In this respect, social science experiments can only establish what people say they would do in a certain situation, or what they would do in a controlled setting — not what they actually would do when confronted with a real situation.

Such experiments are also skewed from the onset, by limiting choices and by subtly applying psychological pressure. In the case of the train car, for instance, why should an individual even be made to feel morally obliged to choose the place of the fatal collision? If the impending accident is the fault of another's negligence for example, or even the result of a freak mishap, why should the bystander be forced to choose which persons to spare when he is neither culpable for any resulting deaths nor equipped to know which lives merit saving?

Similarly, it is a subtle skewing of a noble principle to intimate that if it's good to save five lives instead of just one, then flipping a railroad switch is just as moral an act as shanghaiing a bystander in order to harvest his organs. But again, even this sort of intimation stems from a false comparison based on a false original principle. If, as America's founders insisted, all men are created equal, and if, as the Christian ethos declares, all human life is sacred, then in our roles as individuals we can never truly have the moral right to favor one human life over another. In cases where life-and-death decisions must be made by persons invested with the proper authority, the only just decision is to select what appears to be the least of many evils. And even then, as human beings, we can never be certain of the outcome. The law of unintended consequences tells us this, and writers of science fiction especially have illustrated this truth to the point of satire.[14]

These criticisms as to the credibility of social science in determining what humans really think and how they should live are reflected by others. Lewontin, for instance, takes to task sociologists who study human sexuality. He argues that these scholars exaggerate the reliability of their research methods and consequently present a flawed picture of human behavior. The reason for this, he says, is that human culture can never be accurately rendered from so-called scientific random sampling and that human individuals as a whole cannot be relied upon to reveal the complete truth about their behavior.[15]

Lewontin concludes:

> "There are some things in the world we will never know and many that we will never know exactly. ...By pretending to a kind of knowledge that it cannot achieve, social science can only engender the scorn of natural scientists and the cynicism of humanists."[16]

Postman, as well, disparages the idea that modern sociology delivers newer, more detailed insights regarding human behavior. Social science, he declares, simply repackages common observations into a new kind of narrative — one told with graphs, charts, and other statistical tables. "Thus," he writes, "the purpose of doing this kind of work is essentially didactic and moralistic."[17] We might add that the moral of the story depends on who is doing the telling.

Finally, the argument from social science that human beings can naturally develop good rules to live by is dubious for reasons almost too obvious to mention. As the new atheists point out themselves with wearisome repetition, what people say they believe is good does not necessarily translate into how they live their lives. And even establishing that human beings agree certain kinds of behavior are good is not the same thing as explaining why they are good.

But as we are about to see, the new atheists and other scientific materialists are ill equipped to discuss why certain human actions are good because of a fundamental handicap: According to modern science, it's doubtful whether human beings even have the capacity to choose one behavior over another.

Chapter Ten

No One To Blame

New Atheists Disdain Religious Behavior But Deny Free Will

"The person who asserts his freedom by saying 'I determine what I shall do next,' is speaking of freedom in or from a current situation: the I who thus seems to have an option is the product of a history from which it is not free and which in fact determines what it will do now."
— B.F. Skinner, *About Behaviorism*[1]

"Man is neither a plaything of the gods, nor a marionette suspended on his chromosomes."
— Arthur Koestler, *Janus: A Summing Up*[2]

Are human beings free to choose how to live their lives? Granted that they can even determine that certain actions are evil, can they of their own free will choose the good instead?

These are crucial questions for several reasons. To begin with, as we've already discussed, one of the key pieces of evidence the new atheists present against the existence of God is their bitter disappointment that people, like nature in general, aren't more benevolent. What these unbelievers don't admit is that their argument concerning evil presumes a better world could exist — and that no rational creator (assuming one existed) would choose to design a cosmos as wracked with evil as the one we inhabit. This same logic implies that human beings as well, if they indeed are rational and free, should not choose so often to commit evil when they could be doing good.

And these arguments would truly be damaging to theism if they weren't based on false views of both divine and human nature. But before we consider in detail these particular weaknesses in the new atheists' thesis, it would help first to examine a bit of irony that undermines the entire accusation of human culpability — including the charge of religious hypocrisy. Namely, according to the scientific view held by the new atheists, humans are no more free in their actions than are elm trees and eels.

> **Key Concepts**
>
> The fact that humans often do evil instead of good appalls the new atheists. Yet, they aren't sure humans have the capacity to choose. Consider:
>
> ❖ Some Darwinists claim all apparent choices are merely responses to stimuli.
>
> ❖ Others suggest human will is determined by genetics or other external influences.
>
> ❖ Meanwhile, Christian doctrine insists humans are free, and that evil is the result of abusing that freedom.
>
> ❖ Proof of free will is apparent in the fact that humans are also free to suffer the consequences of their choices.

Behaviorists in Denial

Enquirers tackling one of the most extreme models of human consciousness and will — behaviorism — would do well first to brace themselves for a good deal of intellectual gymnastics. This is because behaviorism, like so many other academic fields, has acquired its own peculiar vocabulary which students must master in order to understand principles the theory propounds. Adding to the usual confusion, unfortunately, is how behaviorism — and related theories of human nature — takes otherwise normal words and assigns them quite different meanings.

Thus warned, we can examine how it is behaviorists insist that the human mind, thoughts, ideas and choices simply don't exist.

Behaviorism makes these claims by working from a now familiar foundation: Darwinism. Starting from the evolutionary idea that man is just an advanced animal, and that animals are nothing more than biological machines, behaviorist proponent John Watson argued in the early twentieth century that animal actions could be described completely in terms of stimulus and response.[3]

Among those who advanced Watson's ideas was B.F. Skinner, probably most remembered for the way he extrapolated research on rats and applied it to theories of human behavior. Rats in Skinner's laboratories developed predictable behaviors after certain actions were consistently rewarded or punished in accordance with carefully documented procedures called "reinforcement." Based on their success in manipulating animals, Skinner and fellow behaviorists argued that human actions, too, were the result of positive or negative reinforcement, and that there was no need to account for nebulous motivations such as "feelings," "beliefs" or "choices," which they disparaged as false "mentalist" constructs.

Skinner writes:

> "Operant behavior is called voluntary, but it is not really uncaused; the cause is simply harder to spot. The critical condition for the apparent exercise of free will is positive reinforcement, as the result of which a person feels free and calls himself free and says he does as he likes or what he *wants* or is *pleased* to do."[Emphasis in the original][4]

This focus on stimuli may strike some as obsessive and simplistic, but behaviorists defended the view as scientific because it treats choices and actions as empirical phenomenon which can be objectively measured. Again, to quote Skinner: "If a behavioristic interpretation of thinking is not all we should like to have, it must be remembered that mental or cognitive explanations are not explanations at all."[5]

With due regard for environmental influences, from random disasters to human kindness, it should be pointed out, however, that simply ignoring intangibles such as emotions and ideas offer no explanations either. However predictably rats in mazes may behave, their actions cannot be equated with human behavior precisely because humans engage in the kind of mental life that no animal is known to display.

It's absurd to think, for instance, that any behaviorist who ever wrote about his theories actually believed that of the thousands of words he individually selected in order to compose a persuasive argument, not one was chosen of his own free will. Was each keystroke made on the basis that some other keystroke (negative reinforcement) would not have delivered the desired accolades and academic recognition (positive reinforcement)? Or more to the point, if there are no ideas — no absolute truths — how do we know Skinner would not have been just as positively reinforced if he had abandoned research into animal behavior and had become instead an itinerant evangelist?

> **Key Quote**
>
> Without a basis for morality, the new atheists have no grounds for excoriating religious belief or for claiming that their own agenda is more rational and humane than one allegedly based on divine revelation.

This bit of hypothetical silliness is akin to asking if Darwinism is just another meme, but it does bring us to the crux of why evolutionary arguments against free will are so problematic. If behaviorism is correct, then God not only doesn't exist, but any thoughts of him are less than fantasy because they are simply nonscientific byproducts of material existence. But then, equally fantastic are concepts of good and evil, as are any moral systems based on these concepts. Thus, it is impossible to describe any behavior as moral or immoral — a situation which poses major difficulties for the new atheists. Because without a basis for morality, they have no grounds for excoriating religious belief or for claiming that their own agenda is more rational and humane than one allegedly based on divine revelation.

But the new atheists do not call themselves behaviorists. They recognize human freedom to make certain choices, and they speak of important ideas which can be grasped by the human mind. But these admissions are, like behaviorism, founded on the principle that all aspects of life must be explained through the lens of Darwinian evolution, and that the will is ultimately just the expression of materialistic cause and effect. In this regard, the new unbelievers put forward a view of human nature that is at best ill-defined, and at worst utterly unhelpful to their aim of marginalizing religious mores.

To begin with, like behaviorists, Dawkins and Dennett also define humans in terms of biological machinery. Employing the sophisticated language of molecular biology, Dennett describes each of us as a compilation of tiny automatons, complex organisms to be sure, but completely lacking any sort of individual essence that might be described as an immaterial soul. As he puts it: "Each of your host cells is a mindless mechanism, a largely autonomous micro-robot. It is no more conscious than your bacterial guests are. Not one of the cells that compose you knows who you are, or cares."[6]

Nevertheless, Dawkins and Dennett declare that humans are in some way more than just the sum of all their robots, in that they can select from among the options that evolution provides for them. These options, they say, are delivered via a variety of means, many of which we've already considered: memes, evolved patterns of behavior, and human culture.

But whether or not human beings are intrinsically capable of rational choice, whether or not they can will themselves to pursue a certain course simply because they think it is the moral thing to do, remains an open question.

Dawkins, for instance, declares that just because our genes are selfish — in the Darwinian sense of struggling against other genes for survival and propagation — that human beings are not necessarily predetermined to be selfish in the moral sense.[7] Dennett as well insists that the behavioral programming in our genes doesn't necessarily dictate our every action, but he also doubts that humans are capable of completely rising above their "brute biological heritage."[8]

This nebulous, quasi-mystical treatment of human will is echoed in another statement by Dawkins, cited by Phillip Johnson in a critique of both meme theory and the concept of the selfish gene. Johnson mocks Dawkins' call for fellow human beings to promote generosity and altruism in order to rebel against the natural selfishness of their genes.[9]

Johnson writes:

> "This is not only absurd but embarrassingly naïve. If human nature is actually constructed by genes whose predominant quality is a ruthless selfishness, then pious lectures advocating qualities like generosity and altruism are probably just another strategy for furthering selfish interests."[10]

Johnson is, of course, being satirical. But his wit is well employed considering the conflicted moral views displayed by other new atheists, who apparently don't appreciate that, as Johnson puts it, "even Dawkins recoils from the logic of his own position."[11]

This conflicted morality we will consider in detail shortly. For now it's enough to note that when it comes to determining how we should live, or even if we're free to try to become better creatures, it seems science has nothing definitive to tell us. For answers to these fundamental questions, then, we will simply have to look beyond the narrow scope of materialistic thinking.

Choice Experiments in Quantum Physics

Not that science is devoid of any positive evidence regarding human consciousness and free will. As it turns out, experiments in quantum physics (a field which we've already noted is more than a little weird) consistently produce outcomes that seem to depend on how scientists choose to conduct their research. These results are not predicated merely on how the experimenter manipulates the apparatus involved, although that does play a part. The astonishing fact is that the behavior of the physical objects scientists study — in this case atoms or sub-atomic particles —

appears to be affected by the very nonphysical intervention of human consciousness. What's more, in certain cases this influence seems to occur instantly over great distances.

> **Key Quote**
>
> The astonishing fact is that the behavior of the physical objects scientists study — in this case atoms or subatomic particles — appears to be affected by the very nonphysical intervention of human consciousness.

Though widely accepted, this principle of observer influence is not necessarily popular. As we have already discussed, most scientists are extremely reticent to explore any phenomenon that defies a purely materialistic explanation. This includes the implications of quantum physics, which physicists Bruce Rosenblum and Fred Kuttner admit makes their compatriots quite uncomfortable. "We're not unsympathetic to the reaction of our colleagues," they write. "Our discipline's encounter with consciousness sometimes embarrasses us as well — particularly when the encounter is claimed to confirm metaphysical philosophies."[12]

Embarrassing or not, add Rosenblum and Kuttner, the facts cannot be denied. Briefly, here they are. The quantum paradox, as the problem behind the observer principle is known, stems from the reality that atomic and subatomic particles display varying, contradictory characteristics. They can be said to exist as discrete, whole objects, and they can be said to exist as nebulous (in the physical sense) waves that spread out over a relatively wide area. Apparently, they exist in both states simultaneously until observed by human beings.

Physicists arrived at this seemingly impossible conclusion through variations of this now-classic experiment: First, a particle is randomly deposited in one of two boxes. If opened one at a time, one box will be found to contain the particle; the other box will be empty. However, if both boxes are opened simultaneously, a screen opposite the boxes will record that it was exposed to a pair of particle waves that emerged from both boxes at the same time. The conclusion? The single particle (or particle wave) exists in both boxes at the same time until an observer looks in one. His choosing to look in one of the boxes causes it to be there. On the other hand, his choosing to open both boxes at the same time allows the single particle to emerge from both boxes at the same time.

Of course, in accounting for this strange outcome the most orthodox explanations steer clear of invoking consciousness or choice. The Copenhagen interpretation does this by treating quantum mechanics as a more-or-less theoretical discipline that allows scientists to describe objects and events not in terms of specific time and place but in terms of the probability of all possible outcomes occurring. Its conclusions are not to be taken in a strict literal sense, though; neither are the bizarre properties of the atomic and sub-atomic microworld assumed to be applicable to the macroworld of living creatures, speeding trains, planets and stars.

The Many Worlds interpretation, on the other hand, employs an argument akin to the multi-universe theory in cosmology in an attempt to resolve the quantum probability paradox. This interpretation argues that all the possible actions of quantum particles — however strange — are real and do occur. It's just that they occur not only in our own reality but in alternate realities as well. Likewise, the various choices an observer could make actually happen in other realities, too, meaning that the apparent free will of the observer in our reality was not really free will, but just one manifestation of all possible choices being acted out — somewhere, maybe.

Ultimately, though, these explanations prove less than satisfactory, not just because they run counter to how humans intuitively view things, but because the consciousness conundrum in quantum physics runs so deep. Consider the difficulties of the uncertainty principle associated with the Copenhagen wave function model. Given the apparent dual nature of atomic objects, it is impossible to pinpoint the position or velocity of any given quantum particle. Further, the more information an observer gains about one of these characteristics, the less he knows about the other. The observer must choose which characteristic he wishes to know more about, and so in some sense he creates the reality he seeks.[13]

> **Key Quote**
>
> The truth is we are free. Human beings know this intuitively. Our lives are spent in a continuous succession of decision-making, with bad decisions often proving more costly and requiring more energy to recover from than the good ones.

Finally, there's the case of particles that were once connected continuing to exhibit a strange affinity even after they become physically separated. Photons, for example, can be examined by physicists for a certain characteristic known as polarization. Twin-state photons, even when forced to fly apart in opposite directions, always display the same orientation despite the fact that as far as we know, no information passes between them. As Rosenblum and Kuttner write, "Quantum theory's explanation for this behavior is [so] hard to believe, it's worth repeating: Twin-state photons do not have a polarization until the polarization of one of them is observed."[14]

The summation of all the evidence from quantum physics, Rosenblum and Kuttner conclude, is quite clear. "Though it is hard to fit free will into a scientific worldview, we cannot ourselves, with any seriousness, doubt it."[15]

What Price Freedom

And why should we doubt? Because the truth is we are free. Human beings know this intuitively. Our lives are spent in a continuous succession of decision-making, with the bad decisions often proving more costly and requiring more energy to recover from than the good ones. In modern society, especially, we're so overwhelmed with choices that the self-help writers and daytime television pop-psychologists never seem to go out of demand.

And oddly enough, the one message that desperate seekers don't hear from their mass-market mentors is the thesis forwarded in one form or another by most modern scientists: That whatever decision you make, you probably didn't have much choice in the matter. Granted, the message may be absent because fatalism is poor starting point for a feel-good, self-empowerment best-seller. But probably the bigger reason why self-helpers don't peddle the determinism of materialist science is that most people recognize it for the bad philosophy it is, just as they recognize that pundits who coolly argue all morality is relative are likely to jettison their equanimity upon receiving an unprovoked punch in the nose.

What's more, it is these common sense objections to materialistic denials of free will that expose as disingenuous the arguments new atheists use to put forth the idea a good God can't exist in an evil universe. Because, as we've noted, if unbelievers argue that free will is an illusion, then they are severely hamstrung in labeling anything anyone chooses to do as evil. And without grounds for defining any action or thing as malevolent, atheism based on the problem of evil becomes an exercise in using one thing you can't prove exists (evil) to explain away another thing you fervently hope doesn't (God).

But there is much more to the philosophy of free will, just as there is more to the new atheists' ire at God for allowing evil to exist. When they lament the presence of so much death and decay in a cosmos said to be created and governed by a benevolent God, what the new unbelievers are really asking is why God doesn't do more to restrain these evils. This is a legitimate question, but one for which there is already a response.

As apologist C.S. Lewis points out, Christianity answers the problem of evil with the doctrine of free will — and the corollary doctrine of how humans abused that divine gift.

Lewis begins by noting the obvious. In any universe such as ours where there is more than one self-conscious agent, there must be a neutral setting against which these self-aware agents can be contrasted.[16] This opens the possibility of harm in two ways. First, this setting may not necessarily be perfectly suited to all agents at the same time, at least, not without occasional manipulation by another being with extraordinary power. In our universe, we would call such instances of manipulation miracles. Secondly, even assuming that the setting is benign with respect to the various self-conscious agents, one of these agents may still use aspects of this setting to threaten or otherwise inflict harm on another agent. Again, the only way to prevent the infliction of harm by one agent upon another is by extraordinary manipulation.

Key Quote

This is not to argue that creatures are free only in a world where evil exists. On the contrary, as another foundational Christian doctrine states, it is only when human beings recognize how desperately they need deliverance from the corruption they live in that they are ready to embrace the salvation that God so freely offers — not only to be absolved of evil but to be removed from its very presence.

Lewis explains:

> "We can, perhaps, conceive of a world in which God corrected the results of this abuse of free will by His creatures at every moment: so that a wooden beam becomes as soft as grass when it was used as a weapon, and the air refused to obey me if I attempted to set up in it sound waves that carry lies or insults. But such a world would be one in which wrong actions would be impossible, and in which, therefore, freedom of will would be void...."[17]

This is not to argue that creatures are free only in a world where evil exists. On the contrary, as another foundational Christian doctrine states, it is only when human beings recognize how desperately they need deliverance from the corruption they live in that they are ready to embrace the salvation God so freely offers — not only to be absolved of evil but to be removed from its very presence. But to return to the salient point: Unless creatures are capable of suffering the consequences of their decisions, free will is meaningless.

A pair of examples from film and television help illustrate this truth:

The 1998 film *The Truman Show*, for example, explores what it would be like for one unwitting person to be the subject of the ultimate television reality show — a round-the-clock broadcast of every aspect of his life. The movie focuses on the main character as he undergoes a not-so-unusual adult crisis, a sort of undefined disappointment as he reassesses his career, his relationships, and the choices he made in building each of those. What he doesn't realize, at least at first, is that all these things were shaped for him, either through the influence of others in the show — all paid actors — or through the machinations of the directors and crew who labor behind the scenes.

The creator and chief director of the show argues that even though Truman is more or less coerced into making certain decisions, that he is being manipulated for his own good and spared the negative consequences that might otherwise occur in an uncontrolled setting. For instance, when Truman shows a romantic interest in a young woman who was not hand-picked for him, and worse, who tries to tell him the truth about his sham existence — she is quickly and conveniently whisked away. Later in the film, after beginning on his own to question the surreal nature of his surroundings, he acts out in various ways — disrupting traffic, skipping work, hamming it up for concealed cameras which no one else dare admit are there. And yet his strange and disruptive behavior initially fails to elicit the rebukes or threats that it would in the real world where conflict between people is the norm. And it is this peculiar kind of manipulation — the constant intervention to warp the world for Truman's benefit and to spare him the repercussions of certain decisions — that ultimately reveals just how enslaved he is. In a setting where even the weather is manufactured to order, all for dramatic effect, Truman cannot possibly be free.

Even the creator and director, who seems at first to be beneficence itself, is willing to kill Truman rather than let him experience free will. In this respect the director is an evil representation of a god who would not permit his creations to fail at anything. Since the show must be skewed toward a peaceful, successful, happy outcome, Truman's own choices must be equally skewed.

Twilight Truth

Another parable of curtailed human freedom comes from an episode of the 1960s television series *The Twilight Zone*, one which imagines a rather unusual fate for an otherwise common criminal named Rocky Valentine. The tale opens with Rocky being gunned down by a policeman, only to awake — apparently unscathed — in a suspiciously hospitable setting. He not only resumes his former habits, gambling for high stakes and making passes at impossibly flashy females, but finds he can't lose. No bet is too chancy, no woman beyond reach. Informed that he has passed into the afterlife, Rocky assumes at first that he has somehow landed in paradise. Not until the pleasures of incessant vice and cheap victories begin to pale does the truth occur to him. He is not in a gambler's Valhalla, but "the other place."[18] He is, in fact, doomed to keep succumbing to his criminal compulsions, but to never again feel even the perverted thrill of having beaten the odds and cheated the system. Without the possibility of ever losing a wager or being scorned by the opposite sex, Rocky's choices to rob, gamble and carouse are meaningless. He's trapped.

There are, of course, problems with both examples that cannot be remedied with simple suspension of disbelief. In the case of *The Truman Show*, for instance, it's hard to accept that a single person could be kept from such an immense secret about himself, especially given his daily interaction with so many other human beings who are in the know. As for *The Twilight Zone's* moral fable, it is not far-fetched to argue that some people are so fond of vice that they would never tire of indulging in it, even if it meant they had to give up some other aspects of personal freedom.

There are further difficulties, as well, with the problem of evil itself, arising primarily from the question of how there came to be so much of it. Many of these difficulties are addressed in the Christian doctrine of man's fall. For now it is enough to recall that this doctrine ultimately blames the woeful state of the Earth and the cosmos on the initiation of man's ongoing rebellion against God, for which the Creator cursed his own creation. And in regard to human behavior, the concept of original sin certainly makes sense. The trouble arises in accounting for certain nastiness in nature, things such as liver flukes, flesh-eating bacteria and corpse flowers. It's possible that the science of evolution may play a role here — not in far-fetched theories of organisms somehow bootstrapping themselves to sentience and sophistication — but in tracing how once benign creatures could become virulent and corrupt.

Laying aside such speculation, however, we can see how the preceding cinematic examples serve well in illustrating our salient point about free will. If God were to constantly intervene, either by manipulating the physical universe or by restraining the consequences of poor human choices, then free will would cease to exist in any meaningful way. It is true that the apparent reluctance of God to intervene does mean that in some sense he is self-limiting. But this self-limitation does not mean that God lacks the power to intervene, only that he chooses not to. It also means that God is not culpable for the ills that arise due to the choices of the free moral agents he created.

Finally, these principles concerning free will and God's self-limitation reveal how absurd a stance the new atheists take concerning the problem of evil. As we've noted, on one hand they disdain the idea that a good deity could exist when corruption abounds. On the other hand, if God did take it upon himself to constantly constrain our actions, our thoughts and desires — even for our own good — that would by no means satisfy the cynical view that permeates the new unbelief. Indeed, the Dawkinses and Dennetts of the world would curse such a god for a tyrant.

Dawkins, in fact, does exactly that when considering another central religious tenet — that God pledges to judge evil-doers and condemn them to eternal punishment in the afterlife. To Dawkins, predictably, the notion that God could consign anyone to timeless retribution amounts to utter cruelty.[19] What Dawkins does not discuss, however, is why he apparently considers it more truly just to allow unrepentant murderers, child abusers, sex slavers, terrorists and other assorted sinners to pass from life to death without ever being held accountable for their misdeeds.

And so the new atheist argument against theism finally hobbles full-circle into self-contradictory nonsense: Religion is false and God a hoax because evil goes unrestrained. But a God who would bind men's wills and sentence them to judgment is too horrible to contemplate.

But in this loathing, the new atheists actually reveal the truth. Free will exists, because otherwise they would not find the possibility of losing this freedom so appalling. Yet humans are unfettered only so long as they are also free to make wrong decisions, which in turn yield malevolent consequences. This is a hard truth, but fundamental to any meaningful understanding of human nature, true justice, and the retribution that true justice requires (and as well, the doctrine of substitutionary atonement so central to Christianity). Free will also explains in one respect why human individuals are equal, in the sense that none of us can be compelled to act in any way or to think anything to which we do not assent. No one can be forced to lie, cheat or steal; there is always an alternative, however unpalatable — at least in the short term. We are free, but our freedom comes at a cost.

As Kenneth Miller puts it: "A believer in the divine accepts that God's love and His gifts of freedom are genuine — so genuine that they include the power to choose evil and, if we wish, to freely send ourselves to hell."[20]

Chapter Eleven

Why Be Good?

Manmade Morality Carries Little Authority

> "Whatever its cause, the manifest phenomenon of *zeitgeist* progression is more than enough to undermine the claim that we need God in order to be good, or to determine what is good."
> — Richard Dawkins, *The God Delusion*[1]

> "There is a universal moral law, as distinct from a moral code, which consists of certain statements of fact about the nature of man; and by behaving in conformity with which, man enjoys his freedom."
> — Dorothy L. Sayers, *The Mind of the Maker*[2]

We've established, then, that evil arises only by the corruption of something that originally was good, and that human beings practice evil by choosing to do what they know is wrong.

This leaves unanswered an even more fundamental question: Where does goodness and our knowledge of it come from? Indeed, Lewis argued that accounting for the origin of benevolence and morality poses a much greater philosophical conundrum for human beings than the problem of evil, because in our corrupted state we can only comprehend shadowy projections of the perfection that God intended for us and our cosmos.[3] We can, as St. Paul wrote, ascertain general moral principles by observing nature and human interaction, but ultimately the knowledge of how God intends for us to live cannot be deduced through the kind of materialistic studies we call science. Lewis explains that although human beings are equipped to recognize natural moral law, "This consciousness is neither a logical, nor an illogical inference from the facts of experience; if we did not bring it to our experience we could not find it there. It is either inexplicable illusion, or else revelation."[4]

Lewis, of course, goes on to insist that natural moral law certainly is no illusion. In his classic work, *The Abolition of Man*, he lays out what he calls the *tao*, a fundamental collection of moral principles that various religions and cultures have esteemed throughout human history. This includes principles such as the law of beneficence, duties to parents, elders and ancestors, duties to posterity, and laws of justice and mercy.

> **Key Concepts**
>
> Human beings have an intuitive sense of fairness, and certainly know to complain when they've been wronged. But where does that sense of goodness come from?
> ❖ Christianity states that a general sense of morality can be derived from nature.
> ❖ This makes sense given that even the best of men can't always agree on what is right.
> ❖ New atheists, on the other hand, fail to offer a convincing foundation for morality.

While identifying this basic moral code, however, Lewis also offers a warning. Anyone who would strive to live a moral life must begin by cultivating a respect for the higher power from whom all benevolence ultimately originates. Conversely, attempts to cast off natural moral law and to craft a new system of right and wrong inevitably result in a few self-proclaimed enlightened ones trying to subjugate the rest of humanity. Lewis writes: "I am very doubtful whether history shows us one example of a man who, having stepped outside traditional morality and attained power, has used that power benevolently."[5]

Everlasting Truth

Lewis' doubts regarding humanity's ability to forge on its own a moral society would seem overly pessimistic if it weren't for the evidence. Unfortunately, history is replete with examples of leaders whose unorthodox moral views yielded rather grim results, the grimmest of which we will shortly discuss. But first it is important to note this: Because it reflects what is fundamentally and inalterably good, moral truth is equally fundamental and inalterable. Attempts to construct systems of lawful and unlawful behavior, consequently, must be founded on absolute moral truth or they are doomed to fail.

As we've already considered, certain aspects of moral truth can be divined through observation of the natural world, revealing what some have called natural law. In the cynicism of our own era, the concept of natural law as morality ordained by God has fallen out of favor. Still, it inspired the men who framed the American Constitution and who produced one of the most equitable and prosperous societies the world has known. As Thomas Jefferson famously insisted, it was the law of nature and nature's God that gave him and his fellow rebels the right to declare themselves members of a new society where government derived its authority from the consent of the governed. Jefferson, then, recognized the very pragmatic truth that consensus is necessary for any group of human beings who wish to rise above anarchy. And yet, Jefferson and his compatriots also wisely noted that in any society built on consensus there must also be erected impediments to prevent any particular group — whether the majority or minority — from abusing those in dissent.

> **Key Quote**
>
> A plurality or even unanimity of opinion concerning how things ought to be done within a certain circle of human beings is not enough to make the prevailing system good and just.

In other words, a plurality or even unanimity of opinion concerning how things ought to be done within a certain circle of human beings is not enough to make the prevailing system good and just. For one thing, people being what they are, dissent is always sure to arise, even in what seems to be the most idyllic setting. And with such dissent, the question again emerges of how to deal fairly with those who refuse to conform. But the greater issue, and one recognized by the American founders, is that consensus regarding customs or laws — however necessary for a civil society — cannot negate the higher morality of divine law. For instance, even if everyone taking part in a survey of hypothetical moral dilemmas agreed it was justifiable to occasionally kill healthy visitors to hospitals in order to harvest their organs, the fundamental truth upon which the very cosmos is built would still declare such an act to be murder.

This is why, despite what the new atheists insist, the idea of godless goodness can never be more than an oxymoron. At the very least, attempts to construct moral codes based solely on human reasoning produce absurd hypocrisies. For evidence of this, all we have to do is examine the new unbelievers in their struggles to derive ethical systems from a starting point in Darwinism.

Muddled Morality

Harris, for instance, approaches what he calls a rational basis for ethics by declaring that "questions of right and wrong are really questions about the happiness and suffering of sentient creatures."[6]

His would-be sagacity fades quickly, however, when he confesses that deciding what constitutes sentience poses problems because "a science of consciousness is still struggling to be born."[7] And as for promoting happiness, Harris grumbles that no one has quite yet determined how to define happiness, let alone judge which forms should take precedence over others.[8]

Harris' own muddled foundation for manmade ethics manifests itself in equally convoluted observations about right and wrong. For example, he disparages the United States for not abolishing the death penalty, a punishment he decries as unfair to criminals since their characters, he claims, are mostly shaped by their genes and environment. We have already noted that this Darwinist argument — that external factors predetermine the characters of criminals, and everyone else — is not new. It failed in its earlier forms to deliver humane reforms to the criminal justice system, a fact we shall explore a little later. At this time it is more pertinent to contrast Harris' indignation over the death penalty with his remarkable statement about how to deal with extremely volatile ideas and the people who espouse them.

"Some propositions," Harris warns, "are so dangerous that it may even be ethical to kill people for believing them."[9]

Though certainly shocking, it should be pointed out that Harris is not completely clear concerning the context of this remark. He does not indicate whether he is merely advocating self-defense, as he does when decrying pacifism in the face of religious terrorism, for instance.[10] And it should be recalled that there are precedents for retribution based on ideas alone. Sedition, for instance — especially the kind which advocates violent insurrection — has at certain times been considered a capital offense by otherwise freedom-loving governments. So Harris is not necessarily calling for the summary execution of individuals who casually affirm certain forbidden notions.

Nevertheless, with typical hypocrisy Harris does not seem to see the dilemma that he has put himself in. He believes wrong thoughts can be capital offenses. And yet, if thinking such thoughts is a crime worthy of death, then how can the death penalty be considered excessive and immoral? How are we to reconcile Harris' disgust for the execution of criminals with his assertion that belief alone could be grounds for killing another human being? How can he, on the one hand, seek to absolve some criminals on the basis that they couldn't help what they were doing, but then suggest that some ideas are so dangerous that the people who hold them cannot be suffered to live? Shouldn't he argue that genes, or some other physiological entity, make people predisposed to believe certain things? Finally, if people can't help how they act or believe, how is it rational to punish them at all for what they advocate or even what they do? Harris can't really say.

Hitchens, as well, seems torn between a desire to excoriate all things religious and to simultaneously excuse — on so-called scientific grounds — behavior traditionally thought of as vicious.

For example, he rightfully condemns the Roman Catholic Church for its shameful role in shielding priests who sexually abuse children. This failure is especially grievous, he says, because the taboo against adult-child sex is the "one subject where moral and ethical authority might be counted as universal and absolute."[11] Even here, Hitchens get his facts wrong, as West points out by citing proponents of "scientifically" based sex education who advocate all sorts of prurient relations between adults and children.[12]

The point is that after railing against abusive priests, Hitchens then criticizes religion for forbidding pleasures — sexual and otherwise — that he says evolution designed human beings to enjoy. He extends this moral duplicity to the point where he praises one of his heroes, Martin Luther King Jr., for allegedly not allowing religious restrictions to stand in the way of his own carnal fulfillment. Hitchens gloats while reporting how King, the night before his assassination, indulged himself with a woman who was not his wife. Hitchens then argues that this somehow makes the Baptist preacher and civil rights leader more of a great human than if he had actually adhered to the personal ethical code he espoused:

> "He [King] spent the remainder of his last evening in orgiastic dissipation, for which I don't blame him. (These things, which of course disturb the faithful, are rather encouraging in that they show that a high moral character is not a precondition for great moral accomplishments.)"[13]

Yet Hitchens warns that humans should take care not to overindulge in the behavior that evolution predisposes them to crave, because this may lead to ills that can undermine civilization itself.[14] But how human beings are supposed to defy their own biology, or to determine the demarcation between benign pleasure and vicious perversion, he — like his fellow unbelievers — does not say.

And then there's Dawkins. Blending blackest gall with astonishing naivete, he condemns religious morals as unhealthy and infantile and then, with no apparent sense of irony, looks to the light of a nebulous secular zeitgeist to lead humanity into a golden age of morality based on reason and consensus. Almost as if he's incapable of considering the ravages of mankind's last one hundred years, or of looking beyond the would-be socialist nirvana of Western Europe, the Oxford don actually tries to revive the tired idealism of humanistic progressivism.

Society is getting better, Dawkins insists, especially in regard to social justice issues such as equality for women and minorities, and intolerance for civilian casualties in war.[15] And this consistent improvement, he argues, has nothing to do with religion. What's more, "over the longer timescale, the progressive trend is unmistakable and it will continue."[16]

Dawkins declares that this improving trend is the result of a consensus among "liberal, enlightened, decent people."[17] These decent folk, presumably atheistic, or at the very least, suspicious of traditional religious mores, have, according to Dawkins, arrived at a basic moral code of their own. He writes: "The majority of us don't cause needless suffering; we believe in free speech and protect it even if we disagree with what is being said; we pay our taxes; we don't cheat; don't kill, don't commit incest, don't do things to others that we would not wish done to us."[18]

Chapter Twelve

EMPTY PROMISES

The Secular Morality Favored By Atheists Is Already Poisoning Society

> "Well, one may be genetically programmed for a certain amount of aggression and greed, and yet also be evolved enough to beware of following every prompting. If we gave in to our every base instinct every time, civilization would have been impossible."
>
> – Christopher Hitchens, *God is Not Great*[1]

> "Whatever may be conceded to the influence of refined education on the minds of peculiar structure, reason and experience both forbid us to expect that national morality can prevail in exclusion of religious principle."
>
> – George Washington, *Farewell Address*[2]

Is society — particularly Western society — really all that much better off for embracing secular ideals? Do the new atheists really represent the forefront of a rising enlightenment, one where reason alone will secure prosperity, civility and tranquility? In a word, no. At the risk of sounding misanthropic, the truth demands that these millennial raptures of Dawkins and his cohorts be exposed for what they are — an utter sham.

Even now, in the twenty-first century, human beings and the civilizations they form remain as they always have been — endlessly hopeful but hopelessly conflicted. Consider: The bounty delivered by the advances of science remains unevenly distributed. Social justice and the rule of law still apply to only a fraction of the global populace, some would say, to a minority. As for the most prosperous and progressive part of the world, particularly the United States and Western Europe, a closer look at how people there actually live reveals lofty claims about moral consensus to be so much wishful thinking. Worse, by their own words and deeds, Dawkins, his fellow unbelievers, and Darwinists at large show their alleged commitment to ideals such as fairness, free speech, and sexual restraint to be equally dubious. In the end their blather about atheism breeding a growing brotherhood of rational do-gooders actually amounts to nothing more than rhetorical cover for the advancement of a specific political agenda — the proliferation of the liberal nanny state. This will become readily apparent as we contrast the specific claims of the new atheists concerning morality and the true state of affairs.

> **Key Concepts**
>
> The new atheists insist that progressive, democratic societies based on secular morality are leading us into a new enlightenment. The facts say otherwise:
>
> ❖ Tax evasion is rampant in Western nations, cheating governments out of billions of dollars.
>
> ❖ Science itself is being undermined by the way the establishment refuses to tolerate dissent from Darwinism.
>
> ❖ In keeping with this intolerance, some new atheists advocate banning parents from teaching their children about religion.
>
> ❖ Increasingly secularized legal codes have resulted in greater societal woes — from various kinds of exploitation, to family breakdown, to macabre methods for destroying pre-natal infants.

Avoiding the Tax Man

In accordance with the rising atheistic moral consensus, Dawkins assures us, most people pay their taxes. This statement is true on its face, though perhaps a bit misleading. Most citizens of the United States and European countries do indeed remit duties required by law to their respective governments, but not necessarily out of some sense of civic obligation. In fact, tax evasion in these nations has become a major issue.

According to the Associated Press, the latest Internal Revenue Service study on the subject revealed that in 2001 the gap between what Americans owed in taxes and what they actually paid amounted to $345 billion. "Of that, $197 billion came from underreporting on individual income tax returns and $88 billion from underreporting by corporations and the self-employed. The rest came from those not filing or not paying the proper amount."[3] More astonishing, the article also cited a Cato Institute researcher stating that the U.S. tax compliance rate of 80 percent is one of the highest in West, "well above some European countries with thriving underground economies."[4]

One tactic favored by wealthier tax evaders is to hide personal income in secretive, foreign financial institutions. Some governments have taken measures to counter this particular offense, but the revenue lost remains immense. Journalists with the *Christian Science Monitor* quoted experts who estimated the total wealth stashed in so-called off-shore accounts amounts to at least $5 trillion.[5] By hiding this money, would-be payers of income tax are believed to have deprived Britain of $40 billion annually. Germany's losses are estimated at $30 billion per year.[6]

The consensus regarding honesty and taxes, then, appears more than a little tenuous. In the best situation — the United States — at least one taxpayer in five cheats the government. Worldwide, according to experts, wealthier taxpayers conspire to cheat their respective governments out of $250 billion annually.[7]

With such obvious resistance to taxation, one might ask why Dawkins emphasizes tax payment as particularly virtuous. The answer probably has to do with the high tax rates required to fund the sort of socialist democracies he and his compatriots seem to favor. In Western Europe especially, where many nations have evolved cradle-to-grave social welfare systems, including state-run schools and comprehensive health care, the trend has been to eschew the advice of religious advocates on how best to administer these services and turn instead to secular, "scientific" experts. Given that such experts include scientists, philosophers and journalists — in other words, people like Dawkins, Dennett and Hitchens — it becomes easy to see why the new atheism ideal is not necessarily libertarian in regard to taxes.

Free Speech?

And yet, if the new atheists see themselves as being among the liberal enlightened helping to guide society into secular bliss, Dawkins pledges that in their envisioned role as counselors they will never be so intolerant as to deny the free speech of dissenters. Unfortunately, if the present record of modern secularists — especially Darwinists — is any indication, Dawkins' promise of tolerance is simply not to be believed.

Consider the uproar among unbelievers when the news broke that Antony Flew had converted to theism. As journalist and close friend Roy Abraham Varghese describes it:

> "Curiously, the response to the AP story from Flew's fellow atheists verged on hysteria. … Inane insults and juvenile caricatures were common in the freethinking blogosphere. The same people who complained about the Inquisition and witches being burned at the stake were now enjoying a little heresy hunting of their own. The advocates of tolerance were themselves not very tolerant."[8]

Then there's the case of Francis Collins, whose 2009 appointment to direct the National Institutes of Health under President Barack Obama did not sit well among pundits with atheistic leanings. True, their objections to his nomination were mainly philosophical given that Collins' scientific and medical credentials, as Harris grudgingly confesses, "are impeccable."[9] Collins holds a doctorate in physical chemistry from Yale and a medical degree from the University of North Carolina. He helped discover the gene for cystic fibrosis and led the Human Genome Project.[10]

Nevertheless, Collins' critics insisted his accomplishments were not enough to offset the one thing about him they feared was sure to handicap his ability to guide the nation's top source of medical research funding. Namely, Collins openly believes in a sovereign, transcendent God who sets the limits of human knowledge.

Journalist Chris Wilson, for instance, fretted over Collins' dual assertions that some things proceed from the mind of God and are ultimately unknowable to human beings, and that many observable, glorious things hint at God's hand in creation. As examples Wilson points out Collins' claim that altruism in human beings cannot be accounted for through materialistic explanations and his belief the Earth's finely tuned, life-favoring habitat is no accident. This last notion, writes Wilson, "is the area where Collins' religion is most in danger of intruding on his science. He believes that it's possible to see evidence of the divine in things like physics equations or patterns of human behavior. … That is to say, he thinks the presence of the divine can be directly observed, even if it cannot be measured and tested."[11]

Presumably, Wilson means that, because of Collins' deference to religious explanations and regard for Christian moral barriers, he may be reluctant to fund or pursue research that challenge these cherished ideals. If so, this implication is confirmed by Harris's rather blunt criticism. Harris writes:

> "As someone who believes that our understanding of human nature can be derived from neuroscience, psychology, cognitive science and behavioral economics, among other things, I am troubled by Dr. Collins' line of thinking. I also believe it would seriously undercut fields like neuroscience and our growing understanding of the human mind."[12]

What Harris and other critics leave unexplained, however, is exactly why an intelligent, experienced scientist such as Collins should necessarily be hamstrung in his professional duties simply because he is a committed Christian. After all, an abiding faith in the God of the Bible did not keep men like Kepler, Newton and Pascal from historic accomplishments. In fact, whether or not his faith might actually make Collins a superior choice for a top government position was a question largely ignored by his critics. Wouldn't a devotion to the Christian virtue of compassion, for example, be considered a positive characteristic for a physician? And wouldn't a regard for St. Paul's injunction to obey secular authorities make for a better civil servant? Again, aside from Harris' sophomoric observation that "few things make thinking like a scientist more difficult than religion,"[13] those opposed to Collins' NIH appointment offered little of substance by way of protest. Still, the very fact they objected reveals Dawkins' claim of support for dissent and free speech to be the sham that it is. Especially when we consider that Collins — as a Darwinist — escaped the tar-and-feathering reserved for those the scientific establishment truly revile.

> **Key Quote**
>
> A few recent cases where individuals were blackballed by the Darwinian establishment shows that the prevailing intolerance is prompted not so much by a reverence for the truth as it is out of fear for the prestige and influence that stands to be lost should the status quo be overturned.

Dangers of Dissent

Since evolution attained monolithic status in Western academia, including state-run primary and secondary schools, Darwinists have become notorious for the way they stifle critics. We've already noted how in general scientists and journalists risk censure and ridicule if they question the soundness of Darwinian theory. A brief examination of a few recent cases where individuals were blackballed by the Darwinian establishment further shows that this intolerance is prompted not so much by a reverence for truth as it is out of fear for the prestige and influence that stands to be lost should the status quo be overturned.

Consider the plight of biologist Richard Sternberg. While working as a research associate at the Smithsonian Institute's Museum of Natural History in 2003, Sternberg edited a scholarly journal published by the museum, *Proceedings of the Biological Society of Washington.* As part of his duties, he supervised the editing and peer review process of an article by Stephen Meyer, which considered the origin of biological information from the point of view of intelligent design.

The repercussions were severe. Sternberg recounts how his colleagues subsequently ostracized him and sought to undermine his career. He endured open hostility at work. His political and religious beliefs were questioned to the point that he was accused of allowing his faith to cloud his judgment and consequently taint the reputation of the institute's journal. According to the *Wall Street Journal,* one senior scientist at the museum sent an email labeling the Meyer article "unscientific garbage."[14] Eventually, Sternberg was turned out of his office and subjected to additional supervision and restrictions.

In his defense, Sternberg writes:

> "In sum, it is clear that I was targeted for retaliation and harassment explicitly because I failed in an unstated requirement for my role as editor of a scientific journal: I was supposed to be a gatekeeper turning away unpopular, controversial, or conceptually challenging explanations of puzzling natural phenomena. Instead, I allowed a scientific article to be published critical of neo-Darwinism, and that was considered an unpardonable heresy."[15]

In a similar situation, astronomer Guillermo Gonzalez, whose book *The Privileged Planet* we have already considered, was denied tenure in 2007 by Iowa State University. Gonzalez admitted the decision surprised him, considering his academic record. He's published dozens of peer-reviewed articles; some have appeared in journals such as *Nature*, *Science* and *Scientific American.* He also co-authored a college astronomy text.[16]

Gonzalez believes he was done in by his advocacy for intelligent design, which he insists he never taught in the college classroom.[17] Nevertheless, his religious faith and dissenting scientific opinions clearly provoked hostility from his fellow academics. According to the Associated Press, two years before his tenure ordeal Gonzalez "was the unnamed target of a petition signed by more than 120 ISU faculty renouncing intelligent design as legitimate science."[18] And though Iowa State administrators initially denied that intelligent design was a factor in Gonzalez being denied tenure, the subsequent release of university documents and emails in compliance with a public records request revealed a much different reality. *World* magazine reported that dozens of these email messages "mock Gonzalez and his ID work, lumping him with 'idiots' and 'religious nutcases.' "[19] The magazine goes on to quote the Iowa State Physics and Astronomy Department chair as declaring: "The fact that Dr. Gonzalez does not understand what constitutes both science and a scientific theory disqualifies him from serving as a science educator."[20]

Darwinists, then, are extremely jealous of the monopoly of ideas they enjoy in science and state-run education. Consequently, it comes as no surprise that the new atheists, despite their lip-service in regard to free speech, are fiercely intolerant of opposing views in broader educational settings — or anywhere, for that matter.

The Hand that Rocks the Cradle

In fact, Dawkins, Dennett and Hitchens make it a point in their respective works to decry the religious education of children as a form of child abuse. They apparently fear that among the general populace there are too few intellects with the resilience of say, a Darwin or even a Dawkins, both of whom managed to endure early religious indoctrination before going on to embrace irreligious materialism as adults. In their attempt to depict religious instruction as a warping influence on young minds, the new atheists cite extreme practices such as female circumcision, as if genital mutilation somehow compares to teaching children that God does not want them to tell lies. These sorts of histrionics would be comical if it weren't for one very unpleasant truth: The new atheists really do wish that they could muzzle parents.

This desire to stifle the child-rearing rights of parents in general — particularly religious ones — stems from a political philosophy that we've already noted, that in a society guided by science, the experts must have the final say. A good example of this philosophy realized is in a social compact already ratified by most of the world (but not the United States), the United Nations Convention on the Rights of the Child. Though perhaps well-intentioned in its main purpose to help rescue children from poverty and tyranny, the Convention has also been used in various attempts to coerce sovereign governments into accepting an extremely liberal view of how best to raise children. This is achieved through a body of experts at the U.N., who periodically review various nations' efforts to conform to the Convention. These experts, or as homeschooling advocate Michael Farris calls them, "tyrants in tweed,"[21] then can insist that legislators enact further reform — demands based on the presumption that indigenous lawmakers, educators and even parents simply aren't capable of determining what's best for their children. Thus nations such as Britain, a historic leader in state-funded education and social welfare reform, can end up being chided by U.N. experts for alleged shortcomings in regard to children's "rights" such as:

■ Failing to comprehensively ban spanking, even by parents;

■ Failing to take into account the wishes of children when parents opt them out of certain periods of instruction in sex education;

■ Failing to spend enough money to ensure all children's "social and cultural" rights are preserved.[22]

And despite resistance in the United States, this statist philosophy regarding child-rearing has nevertheless made its presence felt. Included in the massive Obama-inspired health care bills of 2009, for instance, was a little-noted provision for expanding government oversight of children, from preschoolers all the way to babies still in the womb. That the proposed program went largely overlooked is hardly surprising considering some versions of the health care legislation ran longer than two thousand pages, and, it was said, few legislators took the trouble to read all of them. But that's another story.

Central to our consideration is the broad outline of the program, which called on state governments to train "experts" who would then visit "families with young children and families expecting children" in order to achieve "positive effects" on "child and parenting outcomes." The expertise that parents ostensibly were to receive ran the gamut of family life, incongruously described in bureaucratic terms: "age-appropriate child development," "health and wellness issues," and "parenting practices" — to be modeled by the obliging government agents. Nowhere in the proposed legislation, though, did it state whether the experts were expected to be successful parents themselves; presumably their government training alone would qualify them to dispense advice. All of which raises the very pertinent question: Who inspects the inspectors?

Lastly, and most outrageously, the legislation urged that the home visitation program should provide "activities designed to help parents become full partners in the education of their children."[23] The galling thing about this particular would-be edict is in what it implies — that even before children reach the age when they come under compulsory school attendance laws, the government should still be considered their primary educator. And the government, by its track record, has not shown itself to be a very tolerant schoolmaster.

A federal court in California, for instance, ruled against parents who objected to not having been given the chance to opt their children out of a survey which included sex-related questions the parents considered objectionable. The court essentially declared that parents have no standing in such matters because once they drop their kids off for a day of public schooling, the role — and rights — of parenting automatically passes over to the government.[24] And we've already discussed the secular, materialist worldview upon which public education is based. Why should we expect that preschool home visitation programs should be based on a philosophy any less secular and any more tolerant of religious dissent?

How would the government nannies react, for example, to religious objections regarding vaccinations? Or to strong views regarding the differences between young boys and girls, how they should be taught or even dressed? What about the use of religious materials and practices at all? Again, if the example of how public education regards such things is applied, the answer does not bode well for the individual liberties of parents. After all, in the words of the proposed legislation, the government parenting experts are not to rely on tradition, religious values or even common sense, but "clear evidence-based models." Parents, in other words, are to be taught how to fashion warm, nurturing homes based on all the latest social science reports. There's a comforting thought.

> **Key Quote**
>
> According to proposed legislation from the Obama administration, government "experts" would teach parents how to fashion warm, nurturing homes based on all the latest social science reports. Now there's a comforting thought.

Finally, in all fairness, it must be pointed out that the proposed home visitation programs were defined as voluntary, though in this case "voluntary" may be said to have various shades of meaning. To begin with, the very means by which the programs were to be established and proliferated also created a certain level of political pressure to ensure they met their stated goals. First, officials of individual states were to assess the need for home visitation. The assessment would then be included in a grant proposal to the federal government. Once grant monies were received and programs implemented, special care was to be taken to record the outcome for inclusion in annual reports which ultimately would be presented to Congress. Given just the bureaucratic costs involved in such a process, what are the odds that any of the reports would contain something like the following: "No funds were spent because no families volunteered." Try to remember the last time your local public school officials said they could do as good a job educating kids on far less money than they were currently spending and you'll have your answer.

But the point here is not to link the new atheists directly to any particular U.N. policies or domestic child care legislation. In fact, Dawkins and his compatriots do not specifically mention the Convention on the Rights of the Child, nor any other international protocols. But in their general political views, especially in their attempts to equate religious instruction with child abuse, their sympathy with those who would restrict freedoms in the name of global scientific expertise is quite apparent. This explains their eagerness to stifle the freedom of educators who criticize Darwinism, or to prevent parents from teaching their children the thing they value most: their faith.

As Dennett puts it, in regard to religious education:

> "Here's a proposal, then; as long as parents don't teach their children anything that is likely to close their minds
>
> 1. through fear or hatred
>
> 2. by disabling them from inquiry (by denying them an education, for instance, or keeping them entirely isolated from the world)
>
> then they may teach their children whatever religious doctrines they like."[25]

In other words, freedom of speech should exist to the extent that Dennett and like-minded "experts" allow it to.

Crisis of Carnality

As for sexual mores, Dawkins implies that a rising secular consensus has brought humanity into an improved, more enlightened ethos. Presumably, he means that by discarding the notion that God reserved sex for married heterosexual adult couples, human beings are freer to enjoy themselves sexually without falling into the sort of profligate behavior that puts physical and emotional health at risk. What exactly counts as immoral, or at least, dangerous sexual behavior, however, becomes in the secular view extremely ill-defined. Part of this is because the gifts of science have helped minimize the risks of profligate sex. Science provides medicines to prevent and cure sexually transmitted diseases, contraceptives to protect against unwanted pregnancy, and as a last-ditch defense aimed at preventing sex from achieving its natural outcome — abortion.

This latter point about how science makes sex "safer", and consequently more accessible, provokes hearty amens among the new atheists. Hitchens, for example, envisions a day when "unfettered scientific inquiry" will usher in perfect sexual liberty, when human beings will be able to gratify their libidos without fear, disease, or tyranny "on the sole condition that we banish all religions from the discourse."[26] Harris, as well, suggests that in its liberated status, legally condoned sexual activity will include prostitution.[27]

> **Key Quote**
>
> And has this new progressiveness finally freed human sexuality from fear and tyranny? Has tolerance toward sexual expression at last created venues where consensual carnality can be pursued in health and happiness — to the greater good of society? Hardly.

Again, however, these glib slaverings must be put to the question: How does this utopian view of sexual liberty square with reality? Already there are places on the globe — particularly Western Europe — where sexual license exists in balance with supposed legal safeguards against exploitation and health risks. In the Netherlands and Germany, for instance, prostitution has been considered for many years legitimate free enterprise operating under more or less normal government oversight and restrictions. Homosexuality, as well, has in much of Western Europe attained widespread acceptance. In the Netherlands, for example, homosexual unions are recognized by the state as being on par with traditional heterosexual marriages.

And has this new progressiveness finally freed human sexuality from fear and tyranny? Has tolerance toward sexual expression at last created venues where consensual carnality can be pursued in health and happiness — to the greater good of society? Hardly.

Consider the facts. The Netherlands legalized prostitution in 2000, intending, as Marlise Simons of the *New York Times* put it, "to make the sex trade more transparent and protect women by giving them work permits."[28] But with this legal lenience came unintended consequences. Amsterdam's red-light district has since burgeoned into a sprawling marketplace of deviant pleasures, from marijuana cafes to shops offering pornography in all of its forms. As for the prostitution trade itself, it has become a lure for exactly the kind of international crime Dutch legislators sought to keep out.

As Amsterdam Mayor Job Cohen admitted, "We've realized this is no longer about small-scale entrepreneurs, but that big crime organizations are involved here in trafficking women, drugs, killings and other criminal activities."[29]

This unpleasant realization prompted Amsterdam's government to launch a campaign against vice in the city, with officials employing tougher zoning laws, running tax audits, and buying up property in a bid to close down brothels.[30] Given the history of woes already produced by the Netherlands' benign negligence in regard to sexual mores, however, the new effort is certain to encounter difficulties.

Indeed, the Netherlands' prostitution trade apparently has been infiltrated by foreign leaders of organized crime since at least the mid-1990s, when the country hosted an estimated 30,000 prostitutes. Citing multiple sources, the *Factbook on Global Sexual Exploitation* reports that by 1997 about three-fourths of the prostitutes working in Amsterdam were foreigners, and that about 70 percent of these had no immigration papers, meaning they were probably victims of "trafficking," a euphemism for modern slavery. Most of these foreigners are believed to have been brought to the Netherlands by crime bosses from Central and

Eastern European nations, primarily Ukraine and Russia. Meanwhile, many women trafficked from even poorer countries most likely had children they were forced to leave behind.[31]

The situation is similar in Germany, where a 2002 law turned the long-tolerated sex trade into a mainstream industry regulated by permits, health checkups and taxes. Just a few years later, in anticipation of Germany hosting the 2006 soccer World Cup, proponents of legalized prostitution saw an opportunity for the country's 400,000 sex workers to do a booming trade servicing foreign sports fans. Officials in major cities planned to issue extra prostitution licenses and to set up mobile brothels near soccer stadiums.[32] Meanwhile, to counter criticism that the projected bonanza would actually increase violent crime and human trafficking, both government and private organizations launched soccer-themed campaigns aimed at curbing these specific ills. Among them, the German Football Federation sponsored a campaign to give forced prostitution "the final whistle," declaring somberly in campaign literature the federation's aim to protect the rights of sex workers, "ensure that services are voluntary, and to combat social stigma."[33]

Unfortunately, fatuous slogans and banal public education campaigns do little to ameliorate a practice that thrives on degradation and exploitation. According to officials at the U.S. State Department, for example, "Field research in nine countries concluded that 60 to 75 percent of women in prostitution were raped, 70 to 95 percent were physically assaulted, and 68 percent met the criteria for post-traumatic stress disorder."[34] This pattern of abuse associated with the sex trade applies even to allegedly progressive countries like Germany, where an estimated 140,000 prostitutes were forced into their profession.[35] As for the whole of Western Europe, an estimated half a million women are trafficked there every year to ply the prostitution trade.

Worse, there is evidence that easy access to regulated commercial sex only increases appetites for deviant pleasures. To cite the *Factbook* again: "There seems to be a growing demand in the European Union for more extreme and 'exotic' sexual services, some of which in and by themselves produce extensive health risks."[36] In Romania in the 1990s, for instance, children living on the streets of Bucharest were prime targets for Western European pedophiles. One news report told of dozens of men from Germany, France and Britain renting apartments in Bucharest for the sole purpose of sexually exploiting homeless youth there.[37] And in Germany, where just about any sex act with another adult can be purchased legally, hundreds of thousands of people travel abroad every year as sex tourists, bound for places such as Cambodia, where children as young as six are sold into prostitution.[38]

In fact, demand for deviant sexual practices appears to be fueling a worldwide crisis in child exploitation. The United Nations recently warned that an estimated 1.2 million children are trafficked every year, both as cheap labor and as sex slaves. A single child may be sold for as little as six dollars.[39] Though much child trafficking and sex slaving occurs in developing countries, the United States is not immune to the ills of the trade. One recent report estimated a quarter of a million American youth were "at risk of becoming victims of commercial sexual exploitation."[40] And according to civil rights attorney John Whitehead, traffickers are taking advantage of the porous U.S.-Mexico border to smuggle north some 18,000 sex slaves every year. He writes, "Children as young as 11 are forced into the slavery that will break their spirits and, for many, result in death."[41]

Given the heinous nature of these statistics, and the fact that such exploitation exists only to sate the prurience of paying customers, the argument that society benefits from an irreligious, rational approach to sexual mores suddenly rings rather hollow. So does the weak disclaimer about how, while enjoying their liberated sexuality, people of the new moral consensus surely will retain enough self-restraint to recognize the hazards of extreme carnality, such as the aforementioned practice of incest.

Indeed, a few social activists have already advocated in the name of science to remove the stigma from incest, pedophilia and other sexual taboos. Citing study guides published by members of the Sexual Information and Education Council of the United States, John West catalogues a series of arguments for essentially treating any sort of sexual activity between any participants as more or less normal human behavior. Building primarily on the work of controversial sex researcher Alfred Kinsey, SIECUS advocates sought to establish, as West puts it, "that children are sexual beings from infancy and that preadolescent sexual activities are perfectly natural among children."[42] SEICUS writers consequently advised educators to regard it as normal if youth experimented in homosexuality, or if boys in particular dabbled in fetishism, transvestism, voyeurism or other extreme behavior.[43] Sexual contact between children and adults — even close relations — were characterized as nothing truly harmful. West recounts that SEICUS board members John Gagnon and William Simon "consistently downplayed the negative consequences of child-adult sex, urged leniency for child molesters, and discouraged parents from reporting the sexual abuse of their children to the police."[44]

> **Key Quote**
>
> By deriding traditional marriage and the nuclear family as patriarchal, outmoded, sexually bigoted, religious nonsense — what have you — social liberals also undermined what common sense should have revealed as one of the cornerstones of civil society.

Death to Marriage

As if the campaign to sexualize children at increasingly younger ages weren't disturbing enough, there's another troubling trend to consider. In the successful drive to promote a liberal, secular view of human sexuality, social reformers appear to have precipitated a calamity for which they are unwilling to take responsibility — or even recognize. By deriding traditional marriage and the nuclear family as patriarchal, outmoded, sexually bigoted, religious nonsense — what have you — social liberals also undermined what common sense should have revealed as one of the cornerstones of civil society. As it is, the evidence from countries with the most liberal marriage policies now confirms the importance of the marriage and family model as taught by religious tradition.

Of course the traditional family — a man and his wife bound together for a lifetime, sacrificially bearing and rearing children to conform to their cherished values — is not a perfect societal arrangement. But it is less imperfect than other arrangements for the regulation of intimacy, the minimizing of sexual ills, and the raising of children. Similarly, the evidence from social science in determining the value of alternatives to traditional marriage and family is not all that could be wished for, but as a buttress to longstanding philosophical and religious ideals on the subject, remains quite compelling.

For a case study, we return to the Netherlands, where homosexual marriage was legalized in 2000. The new law capped a lengthy campaign in which traditional heterosexual marriage was depicted, not as the ideal, but as one in a variety of arrangements that could be chosen by adult couples. As a supporting premise, sex was depicted primarily as a source of pleasure to be enjoyed by consenting adults, and not intrinsically linked to child-bearing and parenting. The results, unfortunately, bore out the warnings of conservative opponents to the marriage change. Instead of serving to elevate homosexual unions, the new law accelerated the demise of marriage. With further legislation in 2001 extending parental leave rights and tax benefits to all co-habitating couples — whether married or not — the institution became virtually meaningless.[45]

This outcome no doubt pleases some. As Stanley Kurtz of *The Weekly Standard* points out, "Many of Europe's social scientists and public intellectuals are cultural radicals who hope to see marriage replaced by cohabitation and an expanded welfare state."[46] Their hopes have been realized, certainly. As of April 2004, no more than 10 percent of the Netherlands' homosexuals had taken marriage vows. Heterosexual marriage has declined as well.[47]

Even more alarming are two related trends that are not so easy for even cultural radicals to dismiss. The birthrate in the Netherlands — in keeping with most of Western Europe — has fallen well below the baseline for simply replacing the current population.[48] Meanwhile, the percentage of out-of-wedlock births in the Netherlands has soared.[49] In other words, even as many supposedly liberated Dutch couples choose not to have children, many of the children who are born are destined to be raised in a setting that is not predisposed to mold them into healthy, productive citizens.

This latter assertion is supported by fact. Though it should be a matter of simple logic, research illustrates that children fare best when raised by their own married mothers and fathers. As Barbara Whitehead of *The Atlantic* summarized in the early 1990s: "According to a growing body of social-scientific evidence, children in families disrupted by divorce and out-of-wedlock birth do worse than children in intact families on several measures of well-being."[50] Children in single-parent families are more likely to experience poverty, as well as emotional and behavioral problems. They are at higher risk of physical and sexual abuse. "They are also more likely to drop out of high school, to get pregnant as teenagers, to abuse drugs, and to be in trouble with the law."[51]

These and other troubling truths concerning the state of fundamental aspects of Western, liberal society lead to an equally troubling conclusion. The secular moral consensus, if there is one, hasn't proven superior to the moral principles of religion. Disdaining religious principles may indeed have resulted in greater sexual license and other adult freedoms, but this licentiousness also bears a cost. For one thing, if Europe and the United States are any indication, a cultural emphasis on sex as personal fulfillment tends to fuel apathy toward the often thankless duty of rearing children. Similarly, normalizing alternatives to marriage promotes an outlook regarding adult relationships where instability and infidelity are considered the norm, and where children become mere accoutrements to their parents' serial dalliances with commitment.

Is it any wonder then, that after considering the results of adults indulging in the so-called freedoms of sexual fulfillment and easy-come, easy-go relationships, Barbara Whitehead warns:

> "These new families are not an improvement on the nuclear family, nor are they even just as good, whether you look at outcomes for children or outcomes for society as a whole. In short, far from representing social progress, family change represents a stunning example of social regress."[52]

Humbug

Lastly, we come to Dawkins' first and final words concerning secular morality, which combine to form a sort of godless golden rule meant to guide the newly enlightened in heaping benevolence on their fellow creatures. Most new atheists, he assures us, refrain from cruelty and in their dealings with others bear in mind how they themselves wish to be treated. It's a noble assertion, certainly, but when judged in light of the benefits it has actually brought about amounts to little more than what in Darwin's own day used to be called humbug.

> **Key Quote**
>
> Reformers operating in the name of science — especially science predicated on Darwinian materialism — have often shown they prefer expediency over mercy.

In fact, reformers operating in the name of science — especially science predicated on Darwinian materialism — have often shown they prefer expediency over mercy. In questions of whether scientific advances should be curtailed in order to satisfy religious, or even conservative qualms concerning possible human suffering or degradation — science more often than not has had its own way. This unfortunate practice predictably has led to the exploitation of some of the most marginalized members of society. Worse, it continues to do so.

This explains, for instance, why the new atheists argue so adamantly in defense of abortion rights, even to the point of invoking irrationalities. Dawkins, for example, scoffs at the idea that abortion is in any way cruel, arguing that most human fetuses, whose hearts begin beating less than four weeks from conception,[53] aren't neurologically developed enough to feel pain at the time they are terminated.[54] This argument, that fetuses or other less-than-fully-developed humans somehow merit less consideration and grace, is common among materialists. It is little surprise, then, that the case for permitting the abortion of fetuses could be extrapolated into a so-called scientific basis for the oppression and even murder of other human beings.

Such an argument was made in the early 1970s by no less a personage than Nobel laureate James Watson, co-discoverer of DNA's double helix. Pointing out that many defects in infants are not detected until after they are born, Watson suggested that the legal definition of a living person be altered to allow parents to euthanize unwanted newborns. *Time* magazine quotes Watson as saying, "if a child were not declared alive until three days after birth, then all parents could be allowed the choice that only a few are given under the present system. The doctor could allow the children to die if the parents so chose and save a lot of misery and suffering."[55] One is left to ask: Whose suffering, and just how is cruelty supposed to be absent from these scenarios?

Atheist ethicist Peter Singer has expanded on this idea, suggesting that parents should have up to a month to decide whether or not to euthanize their newborns.[56] Singer's proposal is based on a philosophy that due regard and legal protection should be reserved for the most sentient creatures, ones that display rationality and a will to live. This category, he says, includes healthy adult humans. It also includes mature great apes, but not human infants, especially if they're afflicted with defects such as Down's syndrome. Thus, in Singer's view, it would be more ethical to conduct scientific research on a human lying in a coma than it would to experiment on a caged chimpanzee. Yet like Watson, Singer insists his ethics are shaped purely by "an abhorrence of suffering and cruelty."[57]

A cynic could remark in rebuttal that according to this line of reasoning it might be justifiable to permanently ease the suffering of everyone with failed kidneys, for instance, or all humans whose standardized test scores fall below the intelligence rating for mountain gorillas. Sadly, though, even bitter facetiousness pales when one is confronted with the utter cruelty inflicted upon millions and millions by willful men who operated according to systems of so-called science-based ethics. The worst offenders — Nazis and Stalinists — we will consider shortly. First, let us examine what is perhaps more instructive and shameful: The harm scientific moralists have already inflicted in democracies allegedly formed as bulwarks against oppression and injustice.

In the United States in the decades before World War II, a reform movement that hoped to improve society by reducing or eliminating certain perceived genetic defects advocated for policies not too different from those being forwarded in Nazi Germany. Specifically, supporters of eugenics, the so-called science of breeding better humans, argued that lesser individuals — criminals, the disabled and "feeble-minded" — should be culled from the gene pool. Unlike the Nazis' "final solution," the preferred method for dealing with genetic undesirables in America was either to sequester or to sterilize them. Consequently, as West reports, "By the early 1930s, thirty states had enacted sterilization laws." What's more, "By 1931, more than twelve thousand Americans had been sterilized under various state laws, a figure that rose to sixty thousand by 1958."[58]

These draconian measures did not necessarily improve society, but they did expose ironic flaws in materialistic philosophies for reform. In the realm of criminology, for instance, the view that lawbreakers were merely the product of their genes — and not necessarily responsible for their actions — was touted as a more humane, enlightened basis for restructuring criminal codes. Emphasis shifted from retributive sentencing to rehabilitation, with the aim of reducing crime by curing the genetic defects that supposedly created criminals. Unfortunately, as West points out, this meant that instead of being sentenced to a definite prison term in response for committing a specific crime, convicts could be incarcerated in an asylum for an indefinite period, while therapists attempted various courses of treatment without knowing for certain even how to determine when the convict was "cured" and ready to be released.[59]

Eugenics, then, actually inflicted social injustice on several levels. It imposed an oppressive restriction — sterilization — not necessarily in response to an individual's bad behavior, but because an individual displayed physical or mental characteristics that were, by pseudo-scientific consensus, deemed defective. Indeed, as we've already noted, the eugenicists' materialistic ideology as applied to criminal justice undermined the idea that humans could necessarily do good even if they were able to recognize right from wrong. More reprehensible was the way in which eugenics-inspired sterilization was forced upon individuals whose long-term behavior showed that their alleged inability to function in society was grossly misjudged.

Perhaps the most notorious example involved sisters Carrie and Doris Buck, the product of a mother whom Virginia authorities had labeled mentally unfit. Raped as a teenager, Carrie gave birth to a daughter, whom leading eugenicists in the state feared would become another in a succession of indolent idiots, compelled by heredity into sexual promiscuity. Consequently, in 1925 Carrie Buck was chosen as the subject for the test case on Virginia's recently enacted sterilization laws, a trial that wound its way to the U.S. Supreme Court and inspired the infamous commentary by Justice Oliver Wendell Holmes Jr.: "Three generations of imbeciles are enough."[60]

Once sterilized and released into society, however, Carrie Buck proved both able-minded and productive. According to West, she was described as an avid reader and capable letter-writer. She forged stable relationships, including two marriages, the first ending with her husband's death in 1956, the second terminated by her own death in 1983. She was called a trusted caregiver and competent worker. Yet she claimed, as did her sister Doris, who was also sterilized, that authorities never disclosed the purpose of the operations they were forced to undergo. When Carrie Buck did learn, in her sixties, the true reason why her attempts to conceive children with her husband failed, "she was heartbroken."[61]

This sort of callousness in the cause of science continues today. Whitehead recounts how abortion clinics now provide the raw material for certain research and industries in the form of fetal tissue. Fetal cells have been used, for instance, in developing vaccines and for cosmetic treatment. This in turn has spurred the creation of barbarous abortion techniques aimed at preserving the greatest amount of fetal tissue, techniques similar to so-called "partial-birth abortion" in which babies are partially extracted from the womb so that abortionists can puncture their heads, suck out their brains, and leave their bodies intact.[62]

In a similar vein, West reports research from the past several decades in which macabre experiments were conducted on fetuses that were delivered alive. In various cases, living babies were packed in ice, submerged in saline solution, had organs removed, or had their chests cut open so their beating hearts could be observed. The most repulsive case West recounts deals with a brain metabolism study conducted in Finland on a dozen fetuses. "After the fetuses' hearts stopped beating (but with their brain tissue still alive), their heads were cut off and attached to a pump that circulated a chemical mixture 'through the internal carotid arteries'. "[63]

No Better

Here then, is the problem, as illustrated by the blood of innocents. For all the bravado of the new atheists, their so-called secular progressivism offers no improvement over religious mores in delivering what one Old Testament prophet insisted was God's command to do justly and love mercy.[64] If anything, the case studies we've considered only vindicates C.S. Lewis' warning that godless morality simply paves the way for tyrants.

True, religion also can be subject to misinterpretation, manipulation, and outright fraud. But to invoke Lewis again, at least religious mores provide the more trustworthy claim of being transcendent truth, revealed to man by the Creator himself. In this way religion should free humans from endless debate about what is just, allowing them instead to focus on applying justice. By accepting the mandate to do no murder, for instance, scientists could apply their resources toward truly useful research, instead of vainly rationalizing why attempts to animate the severed heads of pre-natal infants is anything other than grossly deviant.

> **Key Quote**
>
> Here then, is the problem, as illustrated by the blood of innocents. For all the bravado of the new atheists, their so-called secular progressivism offers no improvement over religious mores in delivering what one Old Testament prophet insisted was God's command to do justly and love mercy.

But there is another fundamental issue raised by exposing as phony the new atheists' claim that secular progressivism is alive, well, and propelling society toward a glorious age. It may be true that a person need not be religious — or even a theist — to have an inkling about what is right and wrong to do. But can anyone who utterly rejects God, his revelation and redemption, achieve true goodness?

Hitchens doesn't think so. Or more precisely, he thinks the goodness demanded by the God of the Bible shouldn't even be aspired to because it can't be attained. Specifically, Hitchens derides Christ's teaching that the foundation for benevolence must begin — not in the science lab — but with an all-consuming reverence for the God who made us, and a love for fellow human beings that at least matches the love each of us reserves for ourselves.

Hitchens complains:

> "The order to 'love thy neighbor *as thyself*' is too extreme and too strenuous to be obeyed, as is the hard-to-interpret instruction to love others 'as I have loved you.' Humans are not so constituted as to care for others as much as they love themselves: the thing simply cannot be done (as any intelligent 'creator' would well understand from studying his own design.) [emphasis in original]"[65]

This supposed limit to human capabilities for altruism Hitchens presumably bases on social science or some other modern research. His pessimism is pitiable, though. Because without altruism, the future of humanity appears bleak indeed, especially with the ghosts of last century's science-inspired holocausts still haunting us.

Chapter Thirteen

HISTORY DOESN'T LIE
Atheism's Crimes Contrast Sharply With Religion's Record of Reform

"[Hitler] shared with Stalin the same materialist outlook, based on the nineteenth-century rationalists' certainty that the progress of science would destroy all myths and had already proven Christian doctrine to be an absurdity."

— Alan Bullock, *Hitler and Stalin, Parallel Lives*[1]

"Christianity has not made man secure or happy or even dignified. But it supplies a hope. It is a civilizing agent. It helps to cage the beast. It offers glimpses of real freedom, intimations of a calm and reasonable existence."

— Paul Johnson, *A History of Christianity*[2]

The horrors perpetrated by the twentieth century's supreme atheists, Adolf Hitler and his Soviet counterparts, have unfortunately lost some of their power to shock. Hitler, as we've already noted, has been transformed into a post-modern caricature of evil, a bugbear to invoke whenever one wishes to discredit an idea or a person with whom one disagrees, however unjust the comparison. Nevertheless, the crimes of the Nazis and Soviets bear examining again for important reasons. They represent the logical culmination of godless ideologies akin to the ones we've been examining, and they expose the falseness of the new atheists' claim that society can only benefit from casting off religion in favor of secular, science-based morality.

But first, some important caveats. The new atheists certainly are not fascists, neither are they, as we've already noted, communists. Dennett, in fact, mocks Marxism,[3] and Hitchens admits to having reluctantly rejected Trotskyism,[4] a confession which may explain some of his irrational antipathy toward anything religious.

The new atheists also make the obligatory efforts to deny the obvious historic comparisons between their ideas and those of Nazis and Marxists, though their objections for the most part lack conviction. Dawkins lamely argues there is no reason to suppose the atheism of Stalin and Hitler inspired their brutality.[5] Harris and Hitchens insist that communists don't count as true atheists, because their faith in the power of the proletariat to inspire global revolution simply mimicked religious belief.[6] While not wholly specious, this thesis still does little for the cause of new atheism considering that many critics have similarly characterized its ideology as a secular religion founded upon the dogma of scientific materialism. As for Hitler, Hitchens describes him as a quasi-pagan who attempted to perfect the idea of the totalitarian state which religious thinkers like John Calvin allegedly invented. Hitchens also accuses the Christian establishment of helping enable Hitler's rise to power.[7] These charges, too, bear some merit, though not nearly enough to obfuscate the very real connection between godlessness and Nazi and Soviet atrocities.

> **Key Concepts**
>
> History illustrates the danger of pursuing the ideas favored by the new atheists.
>
> ❖ The Nazi and communist regimes of the 20th century represent the logical culmination of "scientific" atheist ideology. These regimes are responsible for history's worst atrocities.
>
> ❖ The legacy of Christianity, on the other hand, is one of humane reform, reason and hope.

In truth, both Nazism and Marxism were thoroughly atheistic. But there is another aspect to the mass crimes inspired by these ideologies that the new atheists fear to mention: They were justified in the name of reason and science.

To craft his political philosophy and morbid vision of racial destiny, Hitler blended Social Darwinism with the Nietzchean concept of the amoral superman. Citing the Nazi leader's own notorious *Mein Kampf*, historian William Shirer explains: "Hitler saw all life as an eternal struggle and the world as a jungle where the fittest survived and the strongest ruled — a 'world where one creature feeds on the other and where the death of the weaker implies the life of the stronger.' "[8]

Hitler's assertion that only the northern white peoples — descended from the Aryans — were capable of creating culture[9] was not in itself so bizarre considering contemporary Darwinism supported such a racist view. As Lewin points out, anthropologists of the early twentieth century typically considered the so-called Nordic races as the most recently evolved, and thus the most biologically advanced, while the darker skinned peoples of Asia and Africa usually were depicted as "primitives" more akin to apes.[10] What was unusual were the extremes Hitler advocated in order to advance the pure, "culture-creating" race, the Germans, at the expense of the lesser races: Slavs, Poles, Gypsies, and especially — Jews. The people least detestable among these, in Hitler's view, were to be dispossessed, marginalized, gradually worn down to make *Lebensraum*, or living space, for the burgeoning German population. As for the rest, Jews in particular, Hitler's "final solution" could be summed up in a single word: annihilation.

No Restraint

In order to establish industrialized genocide as a national policy, Hitler first had to recruit lieutenants, who, like himself, had completely cast off the restraints of religious morality. Scorning Christianity as weak and degenerate,[11] he sought to craft a new morality informed by his alleged scientific understanding of racial destiny and empowered by his own supreme will.

Hitler could do this because he regarded religion primarily as a means to an end — a natural phenomenon that he could manipulate for his own purposes. Not unlike the new atheists, Hitler recognized in human beings an inherent spiritual yearning, but characterized this as simply another tool to be plied in the struggle for dominance. Indeed, he presaged Darwinists such as Dennett and Dawkins in musing that, for whatever advantage religion might contribute towards individual or corporate survival, it could well be that the belief, ritual and emotion we relate to the supernatural are really evolutionary entities promoting their own survival. Again, in *Mein Kampf*, Hitler writes, "By helping to lift the human beings above the level of mere animal existence, Faith really contributes to consolidate and safeguard its own existence."[12]

Whatever his views on its origins in nature, however, Hitler did not make the mistake of undervaluing religion as mere feeling. He admitted that behind faith lay doctrines which imply fundamental truths. This admission, though, was really just another example of the Fuhrer's phony magnanimity. Because when he finally troubled himself to discuss the implications of religious doctrine, his off-hand treatment of the topic made it clear he did not take very seriously the claim that faith represents truth. Hitler writes:

> "Among these [religious tenets] we may reckon the belief of the immortality of the soul, its future existence in eternity, the belief in the existence of a Higher Being, and so on. But all these ideas, no matter how firmly the individual believes in them, may be critically analyzed by any person and accepted or rejected accordingly…."[13]

God, sin, redemption and eternal judgment — these ideas Hitler could take or leave. What really mattered to him was how to forge a novel worldview — a *Weltanschauung* — that would awaken a "sacrosanct conviction" in a political body of followers.[14] For this reason Hitler felt free to employ the language of faith or to say that he admired certain aspects of Christianity without actually affirming that Christ's teaching offered anything of lasting value. When Hitler praised the Roman Catholic Church, for instance, it was not for upholding truth but for its audacity in adhering to what he considered outmoded values — and audaciously demanding its followers hold to these values as well. As for Christianity in general, Hitler disingenuously argued that it gained ascendancy through intolerance and terror, thus setting a precedent for his own Nazi worldview to conquer via similar means.[15]

So when Hitler did mention deity, or refer to himself as being guided by providence, according to historian Alan Bullock this was mere rhetoric, "a necessary if unconscious projection of his sense of destiny which provided him with both justification and absolution."[16] As for the masses, Hitler employed religious ideas in order to rally them to his cause, speaking of discipline, sacrifice, duty, and even spirituality. But these things were all subordinate to the Nazi party, and subsequently, to him.[17] Ultimately, Hitler answered to no one. Bullock writes:

> "Stalin and Hitler were materialists not only in their dismissal of religion but also in their insensitivity to humanity as well. The only human-beings who existed for them were themselves. The rest of the human race was seen either as instruments with which to accomplish their purposes or as obstacles to be eliminated."[18]

Unsurpassed Inhumanity

As for communism — the cause of more death and suffering than any other ideology — its inherent inhumanity began with its founder, Karl Marx. He presented his vision of the classless society where property was freely shared, not as an ideal to be preferred, but as a scientific inevitability which would be brought about by the culmination of history. The epic cycle of strife between oppressors and the oppressed would finally end when the false economy of capitalism collapsed, empowering workers to rise up and forge the long-awaited paradise. According to Paul Johnson, in this way Marx legitimized violence and brutal social engineering policies in what he considered the most powerful terms — by declaring them scientific principles. Johnson writes:

> "Marx was a child of his time, and Marxism was a characteristic nineteenth-century philosophy in that it claimed to be scientific. 'Scientific' was Marx's strongest expression of approval, which he habitually used to distinguish himself from his many enemies. He and his work were 'scientific'; they were not. He felt he had found a scientific explanation of human behaviour in history akin to Darwin's theory of evolution."[19]

Building on Marx's philosophical approval, the communist leaders of the Bolshevik revolution in Russia put the idea of violence as a tool for social reform into gruesome practice. Consolidating power after the Russian Civil War, Vladimir Lenin built the Soviet Union into a totalitarian state, replete with secret police, slave labor camps, show trials and death squads. Unlike Hitler, his policy for stifling dissent and promoting acceptance of materialistic morality also included the active persecution of the Christian church.[20]

As biographer Dmitri Volkogonov writes, "Lenin's philosophy was designed to separate the 'pure' thinkers from the 'impure', the materialists from the idealists. His aim was to demonstrate that a school of philosophy which accepted the existence of religion could not be scientific."[21]

Persecution

As for the actual practice of rooting out religious belief in Russia, according to historian Richard Pipes the communists approached this task from slightly divergent points of view. Lenin, for all his philosophizing, preferred to stamp out faith with "uncompromising atheism."[22] Other leaders preferred a slightly more sophisticated approach. Not unlike Hitler, they recognized in human beings an intrinsic yearning for religious experience.

This yearning, they argued, could not simply be extinguished — instead the focus of its passion would have to be redirected toward a manufactured religion based on reason. Pipes writes: "Much Bolshevik antireligious activity in the early 1920s followed this method, promoting science as the alternative to religion, and developing a Communist surrogate cult with its own divinities, saints and rituals."[23]

Efforts at winning God-fearing citizens over to atheism went beyond mere arguments and other forms of intellectual persuasion. Religious ideas were not only officially marginalized, they were mocked by regime-sanctioned newspapers, in dramatic productions, festivals and other forms of propaganda. Religious instruction for children was abolished outright. But more than quenching theism and any individual manifestations of personal faith, the Bolshevik regime especially strove to weaken — if not destroy — the cultural and civil influence of one Russian institution in particular, the Orthodox Church. Under the communists, writes Pipes, church property was confiscated and state subsidies for clergy and faith-based institutions ended. Clergy themselves were deprived of civil rights, harassed and imprisoned.[24]

> **Key Quote**
>
> Hitler, as well as some communist leaders, recognized in human beings an intrinsic yearning for religious experience. This yearning, they argued, could not simply be extinguished — instead the focus of its passion would have to be redirected toward a manufactured religion based on reason.

In spite of such determined persecution, the Orthodox Church remained one of the few institutions willing to openly rebuke the communists. In October 1919, the Patriarch Tikhon, who represented the church ruling council in dealings with the state, issued what Pipes calls "the most daring challenge to the new regime that any public figure had the courage to issue."[25] Tikhon declared the Bolsheviks should:

> "celebrate the anniversary of taking power by releasing the imprisoned, by ceasing bloodshed, violence, ruin, constraints on the faith. Turn not to destruction but establish order and legality. Give the people the respite from the fratricidal strife they long for and deserve. Otherwise, all the righteous blood you shed will cry out against you (Luke 11:51) and with the sword will perish you who have taken up the sword (Matthew 26:52)"[26]

Needless to say, this scriptural admonishment failed to sway. If anything, in attempting to at last break the church the Bolsheviks resorted to more extreme measures, including one particularly cynical scheme that bears recounting. In the early 1920s, as famine induced at least in part by communist policies swept throughout much of Russia, government leaders conceived of a way to use the calamity as political leverage against the church.

When Tikhon agreed that from among its treasures the Orthodox Church would donate to famine relief its "nonconsecrated" vessels, communists agitated for public outcry demanding consecrated vessels as well. This posed a dilemma, because Church doctrine declared that allowing implements consecrated for the holiest of ceremonies to be used for secular purposes amounted to sacrilege. In an attempt to resolve the impasse, Pipes writes, "Tikhon offered to raise money equivalent to the value of the Church's consecrated vessels through voluntary subscription and the surrender of additional nonconsecrated vessels, but he was refused."[27]

The vessels eventually were taken by force in a campaign that involved bloodshed. More clergy were arrested and put on trial. But this show of justice and government outrage on behalf of the people was a sham, Pipes declares, because the Bolshevik leaders never intended to use the church treasures for famine relief. The sale of these Russian objects on the international market did realize from $4 million to $10 million, but most — if not all of the money — was spent to advance the communists' political and economic needs.[28]

Stalin's Horrors

This pillaging of the church, however, was only a precursor to the full implementation of communism's "scientific" premises, which were fully realized under Lenin's eventual successor, Joseph Stalin. First, Stalin purged the Communist Party — not necessarily to purify its ideals, but to secure his own power. He launched a brutal campaign to strip the Russian peasants of their independence and to corral them onto huge government-run farms. Millions died during the subsequent famine, a manmade calamity triggered when large quantities of crops and livestock were destroyed either by the peasants themselves or their Soviet overlords. Indifferent to the sufferings of others, Stalin relied increasingly on secret police to ensure his own safety and prevent his enemies — or even his allies — from gaining enough political leverage to ever contradict his will. Soviet Russia under Stalin consequently became a terroristic regime where the slightest dissent, or the paranoid whim of the supreme ruler, could result in anyone from a factory worker to a high party official being denounced and dealt with. These constant arrests and purges also served to keep stocked the Soviet Union's immense archipelago of slave labor camps, which by 1933 swelled to a population of 10 million and never fell below that figure until some time after Stalin's death in 1953.[29]

This, then, is communism's legacy. Under Stalin, during the 1929-36 oppression of the peasants, 10 million died.[30] From political executions and deaths in the labor camps, another 10 million perished.[31] In China, under Mao Zedong, millions died in various purges and catastrophic reform movements, including the farcical Cultural Revolution of the 1960s in which students took to the streets wrecking and killing. The following decade saw Pol Pot of the Khmer Rouge in Cambodia preside over the murder of 1.2 million fellow Cambodians.[32] All told, writes Aikman, the number of men, women and children either murdered outright by communist regimes, or who perished as the result of war, famine or other communist-provoked calamities approaches 100 million.[33]

These appalling figures exceed even those achieved by the Nazi regime, which mercifully lasted fewer than thirteen years. In that time, however, Hitler's followers managed to murder at least 4 million Jews (some researchers put the total closer to 6 million),[34] about half the number living in the areas of Europe occupied by Nazi Germany. Of the 5 million Russian military prisoners captured by German forces in World War II, 2 million of them died, and another million remain unaccounted for.[35] Many of these, no doubt, expired while employed as forced labor for the worker-starved Third Reich. Shirer estimates that by September 1944 the Nazi slave labor force made up of foreign civilians and prisoners of war totaled 10 million.[36] Added to these atrocities were the Gypsies, communists, mentally and physically handicapped, homosexuals, Christian dissidents, hostages and various political undesirables who were murdered either out of principle or expediency.

This short appraisal of Nazism and Marxism does not touch on certain other bizarre manifestations of these godless regimes, such as the cult of personality built up around their respective leaders, or the scripture-like reverence inculcated for guiding texts such as *Mein Kampf*, or Mao's *Little Red Book*. A closer look at these and other quirks, however, would reveal the extent to which followers of Hitler and Marx displayed deep irrationality. And it is this lapse of reason, in addition to the manifold atrocities, that we invoke as a warning against the ideals preached by current agitators who would replace religion with cold human intellect.

As Aikman so profoundly summarizes:

> "Atheistic totalitarian leaders become irrational because they have replaced God with humanity, or more specifically, they have put themselves in the place of God. When a person believes, as Hitchens and [Ludwig]

Feuerback do, that 'man makes religion,' it is the simplest thing to unmake it and, having done so, to erect a Frankenstein-like structure of idolatry that unleashes upon the human race all the wickedness and cruelty of which humanity has shown itself capable."[37]

Positive Force

In contrast to the crimes of atheism, history reveals religion to be the very thing its detractors claim it isn't: a force for reason, social order, charity and hope. This is not to say that religion has always been, or remains today, above reproach. In the case of Western Christianity, the failings of the church have been particularly grievous — the unfortunate result of divine tenets being entrusted to the institutional care of conflicted mortals.

Key Quote

In contrast to the crimes of atheism, history reveals religion to be the very thing its detractors claim it isn't: a force for reason, social order, charity and hope.

In fact, the story of the Western church could be summarized as an ongoing struggle to reform the corruption brought on by its own success. From its genesis as a maligned and persecuted minority, Christianity rose to become an imperial religion, one of the few meaningful vestiges of Rome to survive the collapse of the empire itself. As such, the Roman church in the medieval era became intrinsically linked to civil authority, to the point that it was able to install as state policy the oppression of those who failed to conform to its form of orthodoxy. And so, the ideal of an earthly society guided above all by a regard for the heavenly good devolved into the reality of a worldly institution preoccupied with riches and power. This state of affairs naturally gave rise to some of the blackest incidents in church history.

The Crusades, for example, achieved little if any lasting spiritual good, but did yield atrocities against infidels and

Christians alike. Intolerance for heterodoxy spawned bloody inquisitions and witch-hunts. The religious wars of the Reformation fostered political intrigues and official persecution by both Catholics and Protestants until the Peace of Westphalia established the sanctity of religion within sovereign nations. Further reform was required before individual states began to acknowledge the right of personal conscience in matters of faith.

Roman Reform

Despite these failings, Christianity still can be upheld as the source for much of what is beneficent and just in Western civilization — especially in the realm of social reform. What's more, this foundation for social amelioration can be traced to the earliest of Christ's followers.

Even as they endured persecution, early Christians in the Roman world established a reputation for being charitable. The first churches were quick to heed the admonitions of apostles and other spiritual leaders in establishing relief for widows and orphans, the poor and infirm. Christians extended their relief work to the needy outside the faith. Historian Haskell Miller points out how "In Alexandria and Carthage, for example, through a series of terrible epidemics they stayed on and ministered to the sick and dying after able-bodied pagans had fled in panic."[38] Christianity also elevated the status of women by regarding them as equal to men in God's sight and by opening positions of service to them within the church.[39]

Once the legitimized church eclipsed paganism within the Roman Empire, it began to exhibit the sort of unfortunate contradictions that have marred its history ever since. On the one hand, Christianity continued to inspire reform. As Miller writes:

> "Laws were passed by Constantine and later emperors abolishing crucifixion, encouraging the liberation of slaves, discouraging infanticide and divorce, prohibiting gladiatorial shows and games, and in many other ways

revealing increasing sympathy for the weak and unfortunate."[40]

On the other hand, church leaders also pressed for policies to consolidate the ascendancy of Christianity. Pagan worship was suppressed and pagan temples destroyed or confiscated. Non-Christians were marginalized, to the point that militant heresy eventually became a crime punishable by death.[41] The church, of course, rose to become more or less a state institution, but only just in time to see the regime it had embraced crumble into insignificance.

Ironically, it was the collapse of Rome and the subsequent cultural ebb of the early medieval era that allowed the Christian church to prove its potential as a civilizing force. As Paul Johnson puts it, in Western Europe the church "was the only organized international body left with ideas, theories, a sophisticated hierarchy and advanced cultural technologies, in an empty world which possessed little but tribalism."[42]

In converting many of the Germanic and Celtic tribes that had warred against Rome, Christianity conveyed vestiges of Roman law and social order to barbarians. The church also continued its role as the primary, if not the sole provider, of systematic charity to the needy. According to Johnson, it was the monasteries, with their endowments of land and brain power, that served as models of agriculture and industry in a regressive age.[43] And it was the early church councils which served as the first primitive state parliaments.[44]

It was the church as well which preserved art and scholarship, then laid the groundwork for its renaissance. Monks kept literacy and literature alive through their arduous labors in scriptoriums across Europe; monks in the Celtic tradition produced manuscripts that were also exquisite works of art. Two ninth-century missionary brothers commissioned by the Eastern church, Methodius and Constantine (later called Cyril), invented the first comprehensive written form of Slav, which became the basis of the Cyrillic alphabet.[45] Out of the system of formal education conducted by the church grew the first European universities, a tradition echoed centuries later in colonial North America, where the first colleges on the continent were established as centers of religious training.

Democracy and Faith

After the Reformation, English parliamentarian William Wilberforce showed through his own exemplary efforts what religious individuals can accomplish while working within a democratic system. Through political persuasion Wilberforce and his allies attained first the ban on slave trading within the British Empire and then the abolishment of slavery altogether. What's more, Wilberforce and his fellow evangelicals represented a religious coalition bent on using all the tools of democracy to advocate for widespread improvements in public morality and welfare. As Johnson points out, "Wilberforce subscribed to seventy societies, which embraced, besides purely religious objects, a huge spectrum of human grievance and misery."[46] These organizations addressed the misfortunes of groups ranging from chimney sweeps to widows to refugees.

On the other side of the Atlantic, the Protestant ethic which emphasized the voluntary expression of personal morality in the form of good works led to the foundation of characteristics intrinsic to the American experiment: government recognition of religious liberty coupled with private admission of civic responsibility. Consequently, in the American tradition nearly every institution for social improvement arose out of the expression of religious activism. The first public schools were religious. The first private asylum for the mentally ill was established by the religious Society of Friends. Evangelist George Whitefield helped open one of the first orphanages.[47]

In the later nineteenth and early twentieth century, religious reformers went beyond basic expressions of charity to advocate against social ills manifested by an advanced industrial society. Temperance activist Kate Bushnell investigated and lobbied against forced prostitution in upper Midwest logging and mining communities in the 1880s.[48] Alabama clergyman Edgar Murphy founded a national committee in the early 1900s aimed at reforming child labor laws.[49] Groups of Southern church women banded together in the 1930s to press for an end to mob violence and public lynchings of black Americans.[50] These are but a few examples of the religious influence upon movements to improve prisons, education, health care, women's rights, racial equality and a host of other issues.

Even in the twenty-first century, personal faith and its attendant good works continues to inspire Americans to commit tremendous acts of charity. One especially impressive example of this tradition can be seen in the response to the 2004 tsunami which devastated Indonesia and many other nations in Southeast Asia. As author Arthur Brooks points out, "Six months after the disaster, Americans had donated more than $1.5 billion in cash and gifts."[51] Most of this aid came from private sources.

In fact, the phenomenon of American giving so fascinated Brooks that he made it the focus of extended research. He summarized his findings in a book with a not-so-rhetorical question for a title: *Who Really Cares: The Surprising Truth About Compassionate Conservatism.* His thesis is remarkable: The same ethos associated with the nation's Protestant founders is what compels American charity today. He writes: "Four forces in modern American life are primarily responsible for making people charitable. These forces are religion, skepticism about the government in economic life, strong families, and personal entrepreneurism."[52]

Secular liberalism on the other hand, Brooks insists, tends to deter its adherents from giving of their treasure, time and talents. He declares that liberals are more likely to expect the government to attend to the poor and oppressed through tax-funded programs and social engineering policies such as income redistribution. Consequently, "Secular liberals are poor givers," Brooks writes. "They give away less than a third as much money as religious conservatives, and about half as much as the population in general, despite having higher average incomes than either group."[53]

The disparity between people of faith and secularists when it comes to personal good works is not lost on atheist Roy Hattersley. Reflecting on the faith-based relief given in response to Hurricane Katrina in 2005, Hattersley notes that in general much of the private aid administered to the poor and suffering in both America and his own native England comes from religious groups. Though it galls him to admit it, he agrees, "The correlation is so clear that it is impossible to doubt that faith and charity go hand in hand."[54] Conversely, rejecting God and jettisoning religious mores apparently fails to inspire similar expressions of personal benevolence. Hattersley notes that of all the groups providing relief in the wake of Hurricane Katrina, "Notable by their absence are teams from rationalist societies, free thinkers' clubs and atheists' associations — the sort of people who not only scoff at religion's intellectual absurdity but also regard it as a positive force for evil."[55]

Real Hope

Typically, Hattersley attempts to resolve his quandary by falling back on a quasi-materialistic explanation. He refuses to consider the obvious alternative: That God is indeed real, that his moral laws are as immutable as any physical expression of his creation, and that any good done by human beings who seek his gift of redemption is merely a pale reflection of his divine glory.

History alone should at least point rational inquirers toward this conclusion. Religion, in the way it has been practiced through the centuries, certainly bears its share of guilt in manifold failures to alleviate injustice and suffering. Christians especially, in often forgetting their namesake's declaration that his kingdom is not of this world similarly neglect his admonition to love their neighbors and do good to their persecutors. Still, contrasted to the grim record established by godless philosophies and regimes, the charity performed by people of faith shines like the light of grace which St. John insists the darkness can never quench.

As journalist Marvin Olasky puts it, the legacy established by atheists in power was not calm, rational reform, but the raw exploitation of that power. "The twentieth century," he notes, "was a century of atheists resolving their disputes not by excommunication, but by murdering each other."[56]

On the other hand, Paul Johnson declares, the history of Christianity — however flawed — at least shows that the church helps provide order and comfort for the present, as well as hope for eternity. He writes:

> "We know that Christian insistence on man's potentiality for good is often disappointed; but we are also learning that man's capacity for evil is almost limitless — is limited indeed, by his own expanding reach. Man is imperfect with God. Without God, what is he?"[57]

Chapter Fourteen

FAITH ABIDES

Divine Redemption Remains Humanity's Best Hope

"With the loving mercy of our God, a new day from heaven will dawn upon us. It will shine on those who live in darkness, in the shadow of death. It will guide us into the path of peace."

– Luke 1: 78-79 (New Century Version)

Far from being a malicious lie, faith in God is imperative for the hope of humanity. True, like everything else entrusted to our care, religion is subject to perversion and corruption. But so far nothing better than Christian faith has been introduced for the ultimate benefit of human beings.

Science, certainly, has broadened our understanding of the physical realm and ameliorated our existence by providing benefits from life-saving medicines to frivolous amusements. But science is abused when antagonists toward faith forward it as a prop for secular materialism, particularly when founded on godless Darwinism. As a myth of life origins, evolution is not nearly as sound as its adherents, atheist or otherwise, profess. And science as a whole has not succeeded in overthrowing God. Science, in fact, has done much to reveal the cosmos as a place of astonishing intricacy, the Earth as an extraordinarily hospitable habitat, and life forms as exceedingly complex. In this vein more recent discoveries in cosmology and microbiology have inspired a minority of scientists and thinkers to argue in favor of an intelligent designer. Meanwhile materialistic science continues to struggle to account for the existence of fundamental abstractions, such as human consciousness, free will, and altruism.

For this reason, as well as the grim lessons of history, atheist attempts to replace religious morality with a secular system of ethics amount to gross arrogance. Indeed, before the new atheists presume to inform the world of their new standard of right and wrong, they should be made to answer the following: If people refuse to respect laws purportedly revealed by God himself, why should they be expected to attend to the fatuous moralizing of secularists?

Of course, the genuine reply to such a query was discovered some time ago and finally implemented to maximum effect by last century's Nazis and Soviets. People can be persuaded to a certain extent through reason or emotion, but ultimately the only sure method for carrying an argument is to do so at the point of a bayonet. The strong often lord it over the weak, an axiom which probably explains why Darwinism is historically associated with Nazi and Communist atrocities, and why as a philosophical basis for silencing scientific and religious dissent it remains a grave threat to civil liberties today.

And yet, in all their ignorance regarding history, reason and revelation, the new atheists present their greatest affront in assuming the false right to judge both God and man. They declare their own standard of right and wrong while mocking the fundamental religious doctrines of divine grace and justice. They condemn people of faith for participating in what they consider a vile deception, then heap scorn on the same people for believing themselves sinners and trusting as well that God could and would provide redemption through the blood of Christ. The new atheists compound their hubris by going on to excoriate the idea that God would condemn anyone to eternal punishment — without considering the injustice they indirectly advocate by denying the existence of a divine judge. Indeed, they themselves recognize the human potential for unbridled evil, and the fact that many human beings pass from life to death without ever being held fully accountable for their misdeeds. Being mere mortals, the new atheists certainly can't ensure that justice always prevails. Given this reality, how is it anything other than infantile to suggest that people should be good for goodness sake? Or that it is somehow just to imagine that creatures with the capacity to deceive, hate, torture and kill should be left to their own devices unless someone more powerful forces them to do otherwise or until they finally drift into oblivion?

Ultimately, then, the new atheism crumbles under the weight of its own contradictions but leaves this premise intact: A God with the power to both judge and redeem remains humanity's first and best hope. As St. Paul famously declared to the church in Corinth, Christians of all men are most miserable if Christ is not risen — if there is no divine champion over sin and death. And yet, reasons for belief in a redeemer abound. Our own sense of justice testifies to this. So does our wonder at witnessing order in the cosmos, or in sharing the fruits of the ingenuity and imagination bestowed upon us by our Creator. To our delight, God still grants us beauty for ashes. Science, of course, can never define this truth for us, but it remains something we can most certainly know — by faith.

Chapter One Notes

1. Christopher Hitchens, *God is Not Great: How Religion Poisons Everything*. (New York: Twele, Hachette Book Group USA, 2007), 56.

2. George Johnson, "A Free-for-All on Science and Religion," *The New York Times*, (Nov. 21, 2006).

3. National Commission on Terrorist Attacks upon the United States, *The 9/11 Commission Report: Final Report of the National Commission on Terrorist Attacks upon the United States* (New York: W.W. Norton, 2004), 311, 238-239.

4. Edward J. Larson and Larry Witham, "Leading Scientists Still Reject God," *Nature*, (July 23, 1998). The two researchers cite the results of surveying members of the National Academy of Sciences. Interestingly, mathematicians returned the highest rate of belief in God at 14.3 percent; biologists returned the lowest rate at 5.5 percent.

5. Paul Johnson, *Modern Times: The World from the Twenties to the Eighties*. (New York, Harper & Row, 1983), 5.

6. Larry Pierce, "The Forgotten Archbishop," *Creation* (March 1998), available at: www.answersingenesis.org/creation/v20/i2/archbishop.asp (July 17, 2008).

7. Stephen Jay Gould, *Rocks of Ages: Science and Religion in the Fullness of Life*. (New York: Ballantine, 1999), 5.

8. Ibid, 60.

9. All quotes from the Salk Institute's symposium are from *New York Times* reporter George Johnson's coverage of the event. See note 2.

Chapter Two Notes

1. Sam Harris, *The End of Faith: Religion, Terror and the Future of Reason.* (New York: W.W. Norton, 2005), 221.

2. Gilbert K. Chesterton, *Heretics/Orthodoxy*. (Nashville: Thomas Nelson, 2000), 194.

3. Richard Dawkins, *The God Delusion*. (New York: Houghton Mifflin, 2006), 1.

4. Sam Harris, *The End of Faith*, 15.

5. Daniel C. Dennett, "The Bright Stuff," *The New York Times*, (July 12, 2003) Dennett characterizes "coming out" as a bright as a "liberating" experience. He also envisions brights banding together to exert political influence.

6. Richard Dawkins, *The God Delusion*, 53-55. Here Dawkins berates fellow scientist Gould for allowing that religion might have some merit after all. Dawkins argues that rational people who study the evidence should embrace atheism completely.

7. Ibid, pg. 11.

8. Daniel C. Dennett, *Breaking the Spell: Religion as Natural Phenomenon*. (New York Viking, 2006), 234.

9. Richard Dawkins, *The God Delusion,* 79.

10. Ibid, 191. Dawkins apparently introduced the idea of memes in his seminal 1976 work, *The Selfish Gene*, which was honored in 2006 with a special 30th anniversary edition.

11. Richard Dawkins, *The God Delusion*, 173.

12. Ibid, 174.

13. Daniel Dennett, *Breaking the Spell*, 14.

14. Ibid, pg. 310.

Chapter Three Notes

1. Eldra P. Solomon, Linda R. Berg, Diana W. Martin, *Biology: Seventh Edition*, (Belmont, Calif.: Brooks/Cole-Thomson Learning, 2005) 334.

2. Kenneth R. Miller, *Finding Darwin's God: A Scientist's Search for Common Ground Between God and Evolution*. (New York: Harper Collins, 1999), 174.

3. Richard Dawkins, *The God Delusion*, 116.

4. Neil Postman, *Technopoly: The Surrender of Culture to Technology*. (New York: Vintage, 1993). "The Loving Resistance Fighter" is the title of the final chapter in this book, referring to persons characterized by Postman as those who, among other things, "take the great narratives of religion seriously and ... do not believe that science is the only system of thought capable of producing truth."

5. Kenneth Miller, *Finding Darwin's God,* 184.

6. Ibid, 19.

7. Solomon et al, *Biology,* 337.

8. Ernst Mayr, *This is Biology: The Science of the Living World*. (Cambridge, Mass.: Harvard University Press, 1997), 129.

9. Ibid, 249.

10. Ian G. Barbour, *When Science Meets Religion*. (San Francisco: Harper Collins, 2000), 164.

11. Francis S. Collins, *The Language of God: A Scientist Presents Evidence for Belief*. (New York: Free Press, 2006), 135, 137.

12. Ibid, 77.

13. Ibid, 27.

14. Kenneth Miller, *Finding Darwin's God,* 205, 219

Chapter Four Notes

1. Carl Sagan, *Pale Blue Dot: A Vision of the Human Future in Space* (New York: Random House, 1994), 57.

2. Phillip E. Johnson, "Intelligent Design in Biology: The Current Situation and Future Prospects," *Think: The Journal of The Royal Institute of Philosophy* (February 19, 2007).

3. Antony Flew, "My Pilgrimage from Atheism to Theism: An Exclusive Interview with Former British Atheist Professor Antony Flew" available at http://www.biola.edu/antonyflew/flew-interview.pdf (November 15, 2008).

4. Ibid.

5. Antony Flew with Roy Abraham Varghese, *There is a God: How the World's Most Notorious Atheist Changed His Mind* (New York: Harper Collins, 2007), 96.

6. Ibid, 89.

7. Ibid, 112.

8. Darwin's letter is cited by Phillip E. Johnson in *Darwin on Trial* (Downers Grove, Ill.: Intervarsity Press, 1993), 103.

9. Solomon et al, *Biology,* 389.

10. Michael J. Behe, *The Edge of Evolution: The Search for the Limits of Darwinism* (New York: Free Press, 2007), 155.

11. Solomon et al, *Biology,* 387.

12. Ibid, 385.

13. Stephen C. Meyer, "DNA and the Origin of Life: Information, Specification, and Explanation," *Darwinism, Design, and Public Education* (June 30, 2007), available at http://www.discovery.org/scripts/viewDB/filesDB-download.php?command=download&id=1026 (November 15, 2008).

14. F.H.C. Crick and L.E. Orgel, "Directed Panspermia," *Icarus 19* (1973), available at http://profiles.nlm.nih.gov/sc/B/C/_/scbccp.pdf (November 12, 2008). Crick and Orgel write that they are proposing the idea of directed panspermia because of the problems involved in generating life from nonlife. Some scientists, they point out, "supposed that if life does not evolve from terrestrial nonliving matter nowadays, it may never have done so."

15. Anil Ananthaswamy, "Bugs Could Travel in Comfort aboard Meteorites," *New Scientist,* January 11, 2002. This article reports an experiment by German scientists to confirm whether or not bacteria embedded in meteorites could survive a journey through space.

16. Leonard David, " 'Mars Meteorite's' Link to Life Questioned," available at http://www.space.com/scienceastronomy/solarsystem/mars_meteor_020514.html (May 14, 2002).

17. Solomon et al, *Biology,* pg. 390.

18. William K. Hartmann, *The History of the Earth.* (New York: Workman Publishing, 1991), 83.

19. Solomon et al, *Biology,* 394.

20. Phillip Johnson, *Darwin on Trial*, 33.

21. Ibid, 32.

22. Charles Darwin, *The Origin of Species by Means of Natural Selection or the Preservation of Favored Races in the Struggle for Life* (New York: Mentor, 1958). The dearth of transitional fossils is the first topic Darwin addresses when enumerating "difficulties of the theory" of evolution in Chapter 6.

23. Stephen Jay Gould, "The Evolution of Life on Earth," *Scientific American* (October 1994).

24. Solomon et al, *Biology,* pg. 393.

25. Stephen Jay Gould, "The Evolution of Life on Earth".

26. Francis Hitching, *The Neck of the Giraffe: Darwin, Evolution and the New Biology*. (New York: New American Library, 1982), 116.

27. Solomon et al, *Biology,* 376.

28. Phillip Johnson, *Darwin on Trial*, 61.

Chapter Five Notes

1. Donald Johanson and Maitland Edey, *Lucy: The Beginnings of Humankind* (New York: Simon and Schuster, 1981), 207.

2. Francis Hitching, *The Neck of the Giraffe*, 183.

3. Ibid, 177.

4. Ibid, 180.

5. Donald Johanson and Maitland Edey, *Lucy,* 32.

6. Ibid, 34.

7. Michael D. Lemonick and Andrea Dorfman, "Ardi is a New Piece for the Evolution Puzzle," *Time* (October 1, 2009), available online at http://www/time.com/time/health/article/0,8559,1927200,00.html (October 8., 2009).

8. Ibid.

9. Christopher Joyce, "Move Over, Lucy; Ardi May Be Oldest Human Ancestor," National Public Radio (October 1, 2009), available online at http://www.npr.org/templates/story/story.php?storyId=11387960&ps=cprs (October 8, 2009).

10. Joel Achenbach, "In study of human origins, remains called 'huge'," *The Philadelphia Enquirer* (October 2, 2009).

11. Roger Lewin, *Bones of Contention: Controversies in the Search for Human Origins* (Chicago: University of Chicago Press, 1997), 138.

12. Ibid, 165.

13. William R. Fix, *The Bone Peddlers: Selling Evolution* (New York: Macmillan, 1984) 53.

14. Solomon et al, *Biology,* 412-413.

15. William R. Fix, *Bone Peddlers*, frontispiece.

16. Richard Lewontin, *It Ain't Necessarily So: The Dream of the Human Genome and other Illusions* (New York: New York Review Books, 2000) 60.

17. Solomon et al, *Biology,* 409.

18. Roger Lewin, *Bones of Contention*, 23.

19. Donald Johanson and Maitland Edey, *Lucy,* 22-24.

20. Ibid, 140.

21. Roger Lewin, *Bones of Contention*, 195.

22. Ibid, 194.

23. Donald Johanson and Maitland Edey, *Lucy,* 243.

24. Ibid, 188.

25. Roger Lewin, *Bones of Contention*, 235.

26. William R. Fix, *Bone Peddlers*, 53.

27. Ibid, 129.

28. Hillary Mayell, "Hobbit-Like Human Ancestor Found in Asia," *National Geographic News* (October 27, 2004).

29. John Vidal, "Bones of Contention," *The Guardian* (January 13, 2005), available at http://www.guardian.co.uk/science/2005/jan/13/research.science (November 15, 2008).

30. Ibid

31. Ibid

32. Steve Jones, *The Language of Genes* (New York: Anchor Books, Doubleday, 1993), 102.

Chapter Six Notes

1. Richard Dawkins, *The God Delusion*, 140.

2. Michael J. Behe, *Darwin's Black Box: The Biochemical Challenge to Evolution* (New York: Free Press, 1966), 22.

3. Steve Jones, *The Language of Genes,* 86.

4. Solomon et al, *Biology,* 321-323.

5. Michael J. Behe, *The Edge of Evolution,* 34.

6. Ibid, 75.

7. Ibid, 49.

8. Ibid, 75.

9. Ibid, 57.

10. Ibid, 60-61.

11. Richard Dawkins, *The God Delusion*, 123.

12. Ibid, 132.

13. Wolf-Ekkhard Lonnig, "The Evolution of the Long-Necked Giraffe: What Do We Really Know?" available at http://www.weloennig.de/Giraffe.pdf (part one) and http://www.weloennig.de/GiraffaSecondPartEnglish.pdf (part two).

14. Ibid.

15. Ibid

16. Solomon et al, *Biology,* 360-361.

17. Carl Zimmer, *Evolution: The Triumph of an Idea* (New York: Harper Collins, 2001), 87.

18. Solomon et al, *Biology,* 359.

19. Ibid, 358.

20. Ibid, 347.

21. Jonathan Wells, "Making Sense of Biology: The Evidence for Development by Design," *Signs of Intelligence: Understanding Intelligent Design*, edited by William A. Dembski and James M. Kushiner, (Grand Rapids, Mich.: Brazos Press, 2001), 120.

Chapter Seven Notes

1. Richard Dawkins, *The God Delusion*, 141.

2. Francis Collins, *The Language of God,* 67.

3. Carl Sagan, *Cosmos* (New York: Random House, 1980), 4.

4. Stephen Hawking, *A Brief History of Time* (New York: Bantam Books, 1996), 5.

5. Ibid, 6.

6. Stephen Hawking, *The Universe in a Nutshell* (New York: Bantam, 2001), 21.

7. Stephen Hawking, *A Brief History of Time*, 43-44.

8. Denyse O'Leary, *By Design or by Chance*? (Minneapolis: Augsburg Books, 2004) 15-16.

9. Stephen Hawking, *A Brief History of Time*, 50.

10. Brian Greene, *The Elegant Universe: Superstrings, Hidden Dimensions and the Quest for the Ultimate Theory* (New York: W.W. Norton & Co., 1999), 169.

11. Ibid, 142.

12. Richard Dawkins, *The God Delusion*, 141.

13. Stephen Hawking, *The Universe in a Nutshell*, 86.

14. Brian Greene, *The Elegant Universe*, 368.

15. Guillermo Gonzalez and Jaw W. Richards, *The Privileged Planet: How Our Place in the Cosmos is Designed for Discovery* (Washington, D.C.: Regnery Publishing, 2004), Introduction xv.

16. Ibid, 166-167.

17. Ibid, 137.

18. Ibid, 104.

19. Stephen Hawking, *A Brief History of Time*, 33.

20. Guillermo Gonzalez and Jaw W. Richards, *The Privileged Planet*, 307.

Chapter Eight Notes

1. Richard Dawkins, *The God Delusion*, 114.

2. Russell W. Howell and W. James Bradley, editors, *Mathematics in a Postmodern Age: A Christian Perspective* (Grand Rapids, Mich.: Eerdman's, 2001), 190.

3. Richard Dawkins, *The God Delusion*, 140.

4. Ibid, 114.

5. Ibid, 121.

6. Daniel Dennett, *Breaking the Spell*, 98.

7. Ibid, 128.

8. Ibid, 129.

9. "Endorsing Infanticide?" *Time* (May 28, 1973).

10. Ron Klinger, "What Can be Done about Absentee Fathers?" *USA Today* (July, 1998) "The U.S. is the world leader in families without fathers. From 1960 to 1990, the number of children living only with their mother jumped from 5,100,000 to 15,600,000. Just 27% of American kids live with their biological mother and father."

11. Daniel Dennett, *Breaking the Spell*, 233.

12. Morris Kline, *Mathematics: The Loss of Certainty* (New York: Oxford University Press, 1980), 26.

13. Ibid, 54.

14. Leonard Mlodinow, *Euclid's Window: The Story of Geometry from Parallel Lines to Hyperspace* (New York: Penguin, 2001), 128.

15. Morris Kline, *Mathematics*, 91.

16. Leonard Mlodinow, *Euclid's Window,* 149.

17. Morris Kline, *Mathematics*, 257.

18. Ibid, 233.

19. Ibid, 337-338.

20. Russell W. Howell and James Bradley, *Math in a Postmodern Age*, 141.

21. Ibid, 168-169.

22. Ibid, 73.

23. Ibid, 74.

Chapter Nine Notes

1. Sam Harris, *The End of Faith*, 173.

2. C.S. Lewis, *The Problem of Pain* (New York: Macmillan, 1962), 24.

3. Richard Dawkins, *The God Delusion*, 46, 78.

4. David Aikman, *Weakness of The New Atheism*. This paper was delivered February 19, 2008 as part of the Faith and Reason lecture series at Patrick Henry College in Purcellville, Va.

5. Sam Harris, *The End of Faith*, 172.

6. John G. West, *Darwin Day in America: How Our Politics and Culture Have Been Dehumanized in the Name of Science* (Wilmington, Delaware: ISI Books, 2007), 270.

7. Daniel C. Dennett, *Freedom Evolves* (New York: Viking, 2003), 197.

8. Ibid, 194-195.

9. Ibid, 177.

10. David Aikman, *Weakness of The New Atheism.*

11. Daniel Dennett, *Freedom Evolves*, 303.

12. Richard Dawkins, *The God Delusion*, 222.

13. Ibid, 266.

14. In an episode of the animated television series, *The Simpsons,* "Treehouse of Horror V," Homer Simpson discovers his inadvertent travels back through time alter his family in freakish ways. The episode makes for amusing comparisons with classic science fiction stories such as Ray Bradbury's "A Sound of Thunder."

15. Richard Lewontin, *It Ain't Necessarily So*, 239, 249.

16. Ibid, 254.

17. Neil Postman, *Conscientious Objections: Stirring up Trouble About Language, Technology and Education* (New York: Alfred A. Knopf, 1988), 18.

Chapter Ten Notes

1. B.F. Skinner, *About Behaviorism* (New York: Vintage, 1976), 185.

2. Arthur Koestler, *Janus: A Summing Up* (New York: Random House, 1978) 239.

3. Jennifer Bothamley et al, *Dictionary of Theories* (Detroit: Visible Ink Press, 2002), 50.

4. B.F. Skinner, *About Behaviorism*, 60.

5. Ibid, 114.

6. Daniel Dennett, *Freedom Evolves*, 2.

7. Richard Dawkins, *The God Delusion*, 215.

8. Daniel Dennett, *Freedom Evolves*, 179.

9. Phillip E. Johnson, *The Robot Rebellion of Richard Dawkins: A Review of Unweaving the Rainbow, Christian Research Journal* 22, No.1 (June 14, 2000). Accessed

at http://www.arn.org/docs/johnson/pj_robotrebellion.htm.

10. Ibid.

11. Ibid.

12. Bruce Rosenblum and Fred Kuttner, *Quantum Enigma: Physics Encounters Consciousness* (New York: Oxford University Press, 2006) 154.

13. William R. Fix, *Bone Peddlers*, 216-217. Fix goes on to suggest that human beings are the result of evolution guided by some collective psychic entity. This fantastic speculation aside, his observations about the implications of consciousness in physics are fascinating.

14. Bruce Rosenblum and Fred Kuttner, *Quantum Enigma*, 134.

15. Ibid, 170.

16. C.S. Lewis, *The Problem of Pain,* 32.

17. Ibid, 33.

18. Marc Scott Zicree, *The Twilight Zone Companion* (Beverly Hills, California: Silman-James Press, 1989), 114.

19. Richard Dawkins, *The God Delusion*, 316-320.

20. Kenneth Miller, *Finding Darwin's God*, 291.

Chapter Eleven Notes

1. Richard Dawkins, *The God Delusion*, 272.

2. Dorothy L. Sayers, *The Mind of the Maker* (San Francisco, Harper & Row, 1987), 9.

3. C.S. Lewis, *The Problem of Pain*, 24.
4. Ibid, 22.
5. C.S. Lewis, *The Abolition of Man* (New York: Macmillan, 1955), 78.
6. Sam Harris, *The End of Faith*, 171.
7. Ibid, 174.
8. Ibid, 185.
9. Ibid, 52-53.
10. Ibid, 199.
11. Christopher Hitchens, *God Is Not Great*, 52.
12. John West, *Darwin Day in America*, 308-309.
13. Christopher Hitchens, *God Is Not Great*, 176.
14. Ibid, 214.
15. Richard Dawkins, *The God Delusion*, 270.
16. Ibid, 271.
17. Ibid, 286.
18. Ibid, 262-263.

Chapter Twelve Notes

1. Christopher Hitchens, *God is Not Great*, 214.
2. George Washington's *Farewell Address* is cited from the compilation *Living Documents of American History,* edited

by Henry Steele Commager and published by United States Information Services.

3. Jim Abrams, "Unpaid Taxes Equal $2,680 Per Household," Associated Press (March 5, 2007). Available at http://www.cbsnews.com/stories/2007/03/05/ap/politics/mainD8NLNJTOO.shtml (November 15, 2008).

4. Ibid.

5. Mark Rice-Oxley and Jeffrey White, "In Europe, Widening Probe Targets Tax Haven," *Christian Science Monitor* (March 25, 2008).

6. Ibid.

7. Ibid.

8. Antony Flew with Roy Abraham Varghese, *There is a God,* viii.

9. Sam Harris, "Science is in the Details," *The New York Times* (July 27, 2009), available online at http://www.nytimes.com/2009/07/opinion/27harris.html (August 21, 2009).

10. Gardiner Harris, "Pick to Lead Health Agency Draws Praise and Some Concern," *The New York Times* (July 9, 2009), available online at http://www.nytimes.com/2009/07/09/health/policy/09nih.html (August 21, 2009).

11. Chris Wilson, "Jesus Goes to Bethesda," Slate (July 9, 2009), available online at http://www.slate.com/id/2222562 (August 25, 2009).

12. Sam Harris, "Science is in the Details," *The New York Times.*

13. Ibid.

14. David Klinghoffer, "The Branding of a Heretic: Are Religious Scientists Unwelcome at the Smithsonian?" *Wall Street Journal* (January 28, 2005). Available at http://www.opinionjournal.com/taste/?id=110006220 (November 15, 2008).

15. Rick Sternberg recounted his mistreatment on a personal website, http://www.rsternberg.net (November 15, 2008).

16. Nafeesa Syeed, "Intelligent Design Advocate Denied Tenure at IS," The Associated Press (May 15, 2007). Available at http://www.discovery.org/a/4054 (November 15, 2008).

17. Ibid.

18. Ibid.

19. Mark Bergin, "Smoking Gun: Iowa State Denied Tenure to an ID-supporting Scientist and Then Tried to Cover up Why," *World Magazine* (December 7, 2007). Available at http://www.discovery.org/a/4352 (November 15, 2008).

20. Ibid.

21. Michael P. Farris, "A Deeper Understanding of the Threat of International Law," *The Home School Court Report* (November/December 2007).

22. U.N. Committee on the Rights of the Child, "Concluding Observations of the Committee on the Rights of the Child: United Kingdom of Great Britain and Northern Ireland" (February 15, 1995).

23. All references to the health care bill are from *H.R. 3200: America's Affordable Choices Act of 2009*, introduced in the House of Representatives July 14, 2009.

24. Mike Farris, "A Dangerous Path: Has America Abandoned Parental Rights?" *The Home School Court Report* (July/August 2006)

25. Daniel Dennett, *Breaking the Spell*, 328.

26. Christopher Hitchens, *God is not Great*, 283.

27. Sam Harris, *The End of Faith*, 158.

28. Marlise Simons, "Amsterdam Tries Upscale Fix for Red-Light District Crime," *The New York Times* (February 24, 2008.)

29. Ibid.

30. Ibid.

31. *Factbook on Global Sexual Exploitation: The Netherlands*, available at http://www.uri.edu/artsci/wms/hughes/netherl.htm (November 15, 2008).

32. "Sex Isn't a Spectator Sport: Germany's World Cup Pimping Will Fuel Sex Trafficking," *Christianity Today* (July 1, 2006). Available at http://www.christianitytoday.com/ct/2006/july/2.20/html (November 15, 2008).

33. "Madame Angie and the Prostitution World Cup," *Spiegel* (May 26, 2006). Available at http://www.spiegel.de/international/0,1518,418206,00.html (November 15, 2008).

34. "Sex Isn't a Spectator Sport," *Christianity Today.*

35. "Madame Angie and the Prostitution World Cup," *Spiegel.*

36. *Factbook on Global Sexual Exploitation: Europe*, available at http://www.uri.edu/artsci/wms/hughes/europe.htm (November 15, 2008).

37. *Factbook on Global Sexual Exploitation: Romania*, available at http://www.uri.edu/artsci/wms/hughes/romania.htm (November 15, 2008).

38. Dan Rivers, "Girl, 6, Embodies Cambodia's Sex Industry," CNN (January 26, 2007). Available at http://edition.cnn.com/2007/WORLD/asiapcf/01/23/sex.workers/index.html (November 15, 2008).

39. "Children in India Cheaper than Buffaloes: Report," Reuters (April 3, 2007). Available at http://www.reuters.com/article/worldNews/idUSDEL15186520070403 (November 15, 2008).

40. U.S. Department of Justice, Child Exploitation and Obscenity Section, "Domestic Sex Trafficking of Minors." Available at http://www.usdoj.gov/criminal/ceos/prostitution.html (November 15, 2008).

41. John W. Whitehead, "Sex Trafficking: The Real Immigration Problem" (April 10, 2006). Available at http://www.rutherford.org/articles_db/commentary.asp?record_id=397 (November 15, 2008).

42. John West, *Darwin Day in America*, 295.

43. Ibid, 296.

44. Ibid, 296.

45. Stanley Kurtz, "Going Dutch? Lessons of the Same-sex Marriage Debate in the Netherlands," *The Weekly Standard* (May 31, 2004). Available at http://www.weeklystandard.com/Content/Public/Articles/000/000/004/126qodro.asp (November 15, 2008).

46. Ibid.

47. Joshua Livestro, "Dutch Decline: Losing Interest in Matrimony," *National Review* (June 29, 2004). Available at http://www.nationalreview.com/comment/livestro200406290924.asp (November 15, 2008).

48. "The EU's Baby Blues," BBC News (March 27, 2006). Available at http://news.bbc.uk/go/pr/fr/-/hi/world/europe/4768644.stm (November 15, 2008).

49. Stanley Kurtz, "Going Dutch?" *The Weekly Standard.*

50. Barbara Dafoe Whitehead, "Dan Quayle Was Right," *The Atlantic* (April 1993). Available at http://www.theatlantic.com/politics/family/danquayl.htm (November 15, 2008).

51. Ibid.

52. Ibid.

53. Solomon et al, *Biology,* 975.

54. Richard Dawkins, *The God Delusion*, 293.

55. "Endorsing Infanticide?" *Time* (May 28, 1973). Available at http://www.time.com/time/magazine/article/0,9171,910661,00.html (November 15, 2008).

56. "Living and Dying: Peter Singer Interviewed by Jill Neimark," *Psychology Today* (January 1999).

57. Ibid.

58. John West, *Darwin Day in America*, 87-88.

59. Ibid, 52-23.

60. "Eugenics: Carrie Buck, Virginia's Test Case," The University of Virginia Online Library. Available at

http://www.hsl.virginia.edu/historical/eugenics/3-buckvbell.cfm (November 15, 2008).

61. John West, *Darwin Day in America*, 143.

62. John W. Whitehead, "Aborted Babies are Big Business" (December 3, 2006). Available at http://www.rutherford.org/articles_db/commentary.asp?record_id=385 (November 15, 2008).

63. John West, *Darwin Day in America*, 335-337.

64. Micah 6:8.

65. Christopher Hitchens, *God Is not Great*, 213.

Chapter Thirteen Notes

1. Alan Bullock, *Hitler and Stalin: Parallel Lives* (New York: Alfred A. Knopf, 1992), 386.

2. Paul Johnson, *A History of Christianity* (New York: Borders Books, 2005), 517.

3. Daniel Dennett, *Breaking the Spell*, 337.

4. Christopher Hitchens, *God Is not Great*, 153.

5. Richard Dawkins, *The God Delusion*, 273.

6. Christopher Hitchens, *God Is not Great*, 250.

7. Ibid, 234, 242.

8. William L. Shirer, *The Rise and Fall of the Third Reich: A History of Nazi Germany* (New York: Fawcett Crest, 1960), 86.

9. Robert B. Downs, *Books that Changed the World* (New York: Mentor Books, 1956), 121.

10. Roger Lewin, *Bones of Contention*, 306-307.

11. Louis Untermeyer, *Makers of the Modern World* (New York: Simon and Schuster, 1955), 686.

12. Adolf Hitler, translated by James Murphy, *Mein Kampf* (Los Angeles: Angriff Press, no date given), 213. This edition apparently is a reprint of an English translation first published by Hurst & Blackett, London, 1939.

13. Ibid, 214.

14. Ibid, 210.

15. Ibid, 254.

16. Alan Bullock, *Hitler and Stalin: Parallel Lives*, 386.

17. Ibid, 224.

18. Ibid, 386-87.

19. Paul Johnson, *Intellectuals* (New York: Harper & Row, 1988), 52.

20. Dmitri Volkogonov, translated and edited by Harold Shukman, *Lenin: A New Biography* (New York: The Free Press, 1994), 378.

21. Ibid, 74.

22. Richard Pipes, *Russia Under the Bolshevik Regime* (New York: Vintage, 1995), 339.

23. Ibid, 338.

24. Ibid, 339-340.

25. Ibid, 345.

26. Ibid, 345.

27. Ibid, 348.

28. Ibid, 356.

29. Paul Johnson, *Modern Times*, 274.

30. Ibid, 272.

31. Ibid, 305.

32. David Aikman, *Weaknesses of the New Atheism*.

33. Ibid.

34. William L. Shirer, *The Rise and Fall of the Third Reich,* 978.

35. Ibid, 952.

36. Ibid, 946.

37. David Aikman, *Weaknesses of the New Atheism*.

38. Haskell M. Miller, *Compassion and Community: An Appraisal of the Church's Changing Role in Social Welfare* (New York: Association Press, 1961), 30.

39. Paul Johnson, *A History of Christianity*, 75.

40. Haskell M. Miller, *Compassion and Community,* 31.

41. Paul Hutchinson and Winfred E. Garrison, *Twenty Centuries of Christianity* (New York: Harcourt, Brace and Co., 1959), 63-73.

42. Paul Johnson, *A History of Christianity,* 127.

43. Ibid, 148.

44. Ibid, 136.

45. Ibid, 184-185.

46. Ibid, 371.

47. Haskell M. Miller, *Compassion and Community,* 77-78.

48. Gary A. Haugen, *Good News About Injustice: A Witness of Courage in a Hurting World* (Downers Grove, Illinois: Intervarsity Press, 1999), 55.

49. Ibid, 56-57.

50. Ibid, 58.

51. Arthur C. Brooks, *Who Really Cares: The Surprising Truth About Compassionate Conservatism* (New York: Basic Books, 2006), 117.

52. Ibid, 11.

53. Ibid, 49.

54. Roy Hattersley, "Faith Does Breed Charity: We Atheists Have to Accept that Most Believers are Better Human Beings," *The Guardian* (September 12, 2005). Available at http://www.guardian.co.uk/world/2005/sep/12/religion.uk (November 14, 2008).

55. Ibid.

56. Marvin Olasky, "Atheists vs. Grace." Available at http://townhall.com/Columnists/MarvinOlasky/2007/12/13/atheists_vs_grace (November 14, 2008).

57. Paul Johnson, *A History of Christianity,* 517.

www.ingramcontent.com/pod-product-compliance
Lightning Source LLC
Chambersburg PA
CBHW071637050426
42443CB00026B/510

* 9 7 8 0 6 1 5 4 6 1 2 9 8 *